T0269282

SpringerBriefs in Applied Sciences and Technology

More information about this series at http://www.springer.com/series/8884

Kun Sang Lee · Tae Hong Kim

Integrative Understanding of Shale Gas Reservoirs

 Springer

Kun Sang Lee
Department of Earth Resources
and Environmental Engineering
Hanyang University
Seoul
Republic of Korea

Tae Hong Kim
Department of Earth Resources
and Environmental Engineering
Hanyang University
Seoul
Republic of Korea

ISSN 2191-530X   ISSN 2191-5318   (electronic)
SpringerBriefs in Applied Sciences and Technology
ISBN 978-3-319-29295-3   ISBN 978-3-319-29296-0   (eBook)
DOI 10.1007/978-3-319-29296-0

Library of Congress Control Number: 2016930276

Printed on acid-free paper

This Springer imprint is published by SpringerNature
The registered company is Springer International Publishing AG Switzerland

Contents

Nomenclature

a	Effective cross-sectional area of one gas molecule (Eq. 2.9)
a	Constant determined by experiments based on proppant type (Eq. 2.32)
a	Experimental coefficient (Eq. 2.36)
a_{Dng}	Intercept of Duong model
a_{LGM}	Constant of logistic growth model
A	Empirical fitting constant (Eq. 2.23)
A	Half of total matrix surface area draining into fracture system (Eq. 4.38)
A	Area being drained (Eq. 4.56)
A_{SRV}	Area of total matrix surface area draining into fracture system
b	Klinkenberg parameters
b	Constant determined by experiments based on proppant type (Eq. 2.32)
b	Experimental coefficient (Eq. 2.37)
b	Derivative of loss-ratio, b-parameter (Eq. 4.19)
b	Intercept of square root time plot (Eq. 4.42)
b_k	Dynamic slippage coefficient
b'_{pss}	y-intercept of normalized PSS equation for gas (inverse productivity index)
B	Formation volume factor
B	Empirical fitting constant (Eq. 2.23)
B_i	Parameter for Langmuir isotherm relation
c	Experimental coefficient
c_f	Formation compressibility
c_g	Gas compressibility
c_t	Total compressibility
$\bar{c}_t$	Total compressibility at average reservoir pressure
C	Constant related to net heat of adsorption
C_A	Dietz shape factor
C_L	Combined leak-off coefficient
$\mathbf{C}$	Tangential stiffness tensor
d	Pore diameter
d	Experimental coefficient (Eq. 2.39)

d_{tube}	Diameter of tube
D	Depth
D	Decline parameter, D-parameter (Eq. 4.19)
$\frac{1}{D}$	Loss-ratio
D_i	Arps' initial decline rate of hyperbolic model
D_i	Diffusion coefficient of component i in the mixture (Eq. 5.4)
D_{ij}	Binary diffusion coefficient between component i and j in the mixture
D_k	Knudsen diffusion coefficient
D_1	Decline parameter intercept at day 1 $(t = 1)$
D_∞	Decline parameter at infinite time $(t = \infty)$
E	Young's modulus
E_1	Heat of adsorption for the first layer
E_L	Heat of adsorption for the second and higher layers
f_{iw}	Fugacity of component i in the aqueous phase
F	Slip coefficient
F_L	Nolte after closure linear time function
F_R	Nolte after closure radial time function
$\mathbf{F}$	Body force
g	Intermediate variable
G	G-function
G_c	G-function at closure
G_i	Original gas in place
G_p	Gas cumulative
h	Net pay thickness
h_F	Hydraulic fracture height
h_{slit}	Height of the rectangular slit
H_i	Henry's constant of component
H_i^*	Henry's constant for component i at reference pressure p^*
J	Mass flux or molar flux
k_{app}	Apparent permeability
k_B	Boltzmann constant
k_D	Darcy permeability or liquid permeability
k_{eff}	Effective permeability after including effect of sorption
k_r	Relative permeability
$k(p_{avg})$	Gas permeability at mean pressure (p_{avg})
K	Carrying capacity
K_n	Knudsen number
L	Length of media
L	Fracture spacing (Eq. 4.40)
m_{Dng}	Slope of Duong model
m_L	Slope of linear flow in after closure analysis plot
m_R	Slope of radial flow in after closure analysis plot
m_{sqr}	Slope of straight line during linear flow period in square root time plot
$m(p_{wf})$	Flowing bottomhole pseudopressure

$\varDelta m$	Pseudopressure drawdown with respect to initial pseudopressure
M	Molar mass
n	Maximum number of adsorption layers (Eq. 2.10)
n	Exponent (Eq. 4.25)
n	One-half or one-quarter for linear flow or bilinear flow (Eq. 4.29)
n_{LGM}	Exponent of logistic growth model
N	Avogadro constant (number of molecules in one mole, 6.023×10^{23})
N_i	Moles of component i per unit of grid block volume
$N_{n_c + 1}$	Moles of water per unit of grid block volume
p	Pressure
p_{avg}	Mean pressure
p_{cog}	Oil-gas capillary pressure
p_{cwo}	Water-oil capillary pressure
p_L	Langmuir pressure
p_o	Saturation pressure of gas
p_{wD}	Dimensionless well flowing pressure
p_{wf}	Well flowing pressure
p_z	Net fracture extension pressure above closure pressure or $p_z = p_{ISI} - p_c$
p_1	Upstream pressure
p_2	Downstream pressure
p^*	Reference pressure
q	Flow rate
q_{sc}	Rate at standard condition
q_1	Flow rate at day 1
Q	Cumulative production
r	Pore radius
r_{avg}	Local average pore radius
r_m	Normalized molecular radius size
r_p	Storage ratio
r_w	Well radius
R	Gas constant
s'	Apparent skin
S	Specific surface area
S_{wr}	Residual water saturation
t	Time
t_c	Closure time (Eq. 4.8)
t_c	Material balance time (Eq. 4.45)
t_{ca}	Material balance pseudotime
t_D	Dimensionless time
t_{inj}	Injection time
t_p	Time to end of injection
$\varDelta t$	Time step
T	Temperature
T_j	Transmissibility of phase j

$\mathbf{u}$	Displacement vector
v	Superficial velocity
V	Gas volume of adsorption at pressure p (Eq. 2.5)
V	Grid block volume (Eq. 3.1)
V_{inj}	Injection volume
V_L	Langmuir volume or maximum gas volume of adsorption at infinite pressure
V_m	Maximum adsorption gas volume when entire adsorbent surface is being covered with a complete monomolecular layer
$\bar{V}_i$	Partial molar volume of component i
w	Width
x	Fracture length (Eq. 4.39)
x	Horizontal well length (Eq. 4.40)
x_f	Fracture half length
x_F	Hydraulic fracture half length
y	Stimulated reservoir width
y_i	Mole fraction of component i
y_{ij}	Mole fraction of component i in phase j
Z	Compressibility factor

Greek Letters

α	Biot's constant
α	Tangential momentum accommodation coefficient (Eq. 2.15)
α	Parameter characteristic of system geometry (Eq. 4.59)
α_r	Dimensionless rarefication coefficient
β	Non-Darcy flow coefficient
γ	Exponential of Euler's constant, 1.781 or $e^{0.5772}$
γ_j	Gradient of phase j
δ	Collision diameter of gas molecule
δ_r	Ratio of normalized molecular radius size, r_m, with respect to local average pore radius, r_{avg}
ε	Strain tensor
η	Fluid efficiency
η	Thermo-elastic constant (Eq. 3.15)
λ	Mean free path
μ	Viscosity
μ_f	Mini-frac fluid viscosity
$\bar{\mu}_g$	Gas viscosity at average reservoir pressure
ρ	Molar density of diffusing mixture
ρ_{avg}	Average density
ρ_r	Reduced density
σ	Transfer coefficient (Eq. 3.7)
$\boldsymbol{\sigma}$	Total stress tensor

σ'	Effective stress
τ	Tortuosity
τ_{iomf}	Matrix-fracture transfer in the oil phases for component i
τ_{igmf}	Matrix-fracture transfer in the gas phases for component i
τ_{wmf}	Matrix-fracture transfer for water
τ_{SEPD}	Characteristic time parameter for stretched exponential production decline model
ϕ	Porosity
ψ	Material balance equation
ω	Dimensionless storativity
ω_i	Moles of adsorbed component i per unit mass of rock
$\omega_{i,max}$	Maximum moles of adsorbed component i per unit mass of rock

Superscripts

n	Old time level
$n+1$	New time level
s	Old or new time level

Subscripts

b	Bulk properties
f	Natural fracture
F	Hydraulic fracture
g	Gas phase
i	Initial state
i	Component index
j	Phase index
m	Matrix
$n_c + 1$	Water component
o	Oil phase
w	Water phase

Chapter 1
Introduction

1.1 Unconventional Gas

The hydrocarbon sources from conventional reservoirs are decreasing rapidly. Because global energy consumption is increasing steadily at the same time, conventional reserves alone cannot meet the growing demand. According to EIA Annual Energy Outlook 2015 (EIA 2015a, b), total primary energy consumption will grow by 8.6 quadrillion Btu (8.9 %) from 97.1 quadrillion Btu in 2013 to 105.7 quadrillion Btu in 2040. There is a pressing need for alternative sources of hydrocarbon energy resources. From technical and economic points of view, the expensive sustainable and renewable energy sources cannot compete with the relatively cheap nonrenewable fossil fuels. Therefore, the immediate alternatives for conventional hydrocarbon would be found in unconventional oil and gas resources. As shown in Fig. 1.1a, these unconventional resources come in many forms and include tight gas, shale gas, coal bed methane (CBM), tight oil, shale oil, and oil shale. Figure 1.1b shows worldwide gas resource pyramids that include the general characteristics and global endowment for each resource (Aguilera et al. 2008; Aguilera 2014). Endowment is the summation of cumulative gas production, reserves, and undiscovered gas. Figure 1.1b shows that the total natural gas endowment, excluding gas hydrates, is approximately 68,000 trillion cubic feet (Tcf) and about 70 % of it is estimated in tight and shale gas.

Decades ago, geologists knew there were vast natural gas resources locked in shale rock deep beneath the earth's surface over much of North America. However, it has been remained as a hard-to-produce resource for a long time. As exploration and production companies use special drilling and formation stimulation (e.g., hydraulic fracturing) techniques to make shale reservoir production economically viable, shale gas has been the focus of gas exploration and production in the United

© The Author(s) 2016
K.S. Lee and T.H. Kim, *Integrative Understanding of Shale Gas Reservoirs*,
SpringerBriefs in Applied Sciences and Technology,
DOI 10.1007/978-3-319-29296-0_1

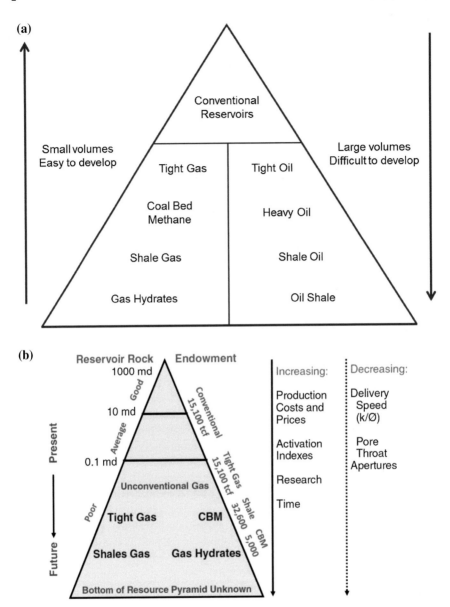

Fig. 1.1 World resource pyramid of **a** hydrocarbon and **b** gas with estimated endowment (Aguilera 2014)

States and in other countries. Based on a recent EIA report (EIA 2013a), there is an estimated 7299 Tcf of technically recoverable shale gas resource to be found in some 95 basins in 41 countries.

As source rocks for most oil and gas deposits, technically recoverable (although not necessarily economically recoverable) gas shale is abundant across the globe. It is also located in a very wide range of geographical regions, and in many of the nations with the highest energy consumption. For certain nations, shale gas, therefore, has the potential to reduce energy prices and dependence on other nations, hence impact on both the political and economic outlook. However, the prospects for and significance of shale gas are greater where there is lack of existing conventional gas production, where there is a lack of existing conventional gas production, where there is proximity to demand (i.e., population), and where some form of existing gas distribution infrastructure exists (Rezaee 2015).

Shale gas resources have received great attention because of their potential to supply the world with an immense amount of energy and the depletion of conventional reservoirs. Shale gas reservoirs have some features dissimilar to those of conventional reservoirs. Typically, shale gas reservoirs include conductive natural fractures that substantially influence well performance. Because shale gas reservoirs have narrow thickness and infinite lateral extension, horizontal wells are usually applied to increase production by augmenting the contact area of the wellbore and the pay zone. Shale matrix has extremely low permeability. To exploit ultra-low permeability reservoirs, hydraulic fracturing technology has been proved to be an effective means. Hydraulic fracturing induces fractures with enormously high permeability and makes fracture networks around the wellbore. Shale gas is stored in both free gas and adsorbed gas. Therefore, shale gas reservoirs show a long period of changing behavior and intricate flow regimes, which makes understanding the pressure behavior of shale gas reservoirs important.

Following notable successes in shale gas production in the USA, to the point where that country now produces more shale gas than gas from the conventional sources, other countries are pursuing the same course. Even so, in order to be successful in the exploration and the development of shale gas plays, a vast knowledge of the shales is required. The aim of this book is to provide some guidance on the major factors involved in evaluation shale gas reservoirs.

1.2 Overview of Shale Gas Reservoir

Due to its environmental friendliness, natural gas has played a prominent role from the late 20th century. Currently, a large portion of the natural gas comes from unconventional sources (e.g., shale gas, tight gas, coal bed methane, and, soon, gas hydrates). Unconventional gas reservoirs are loosely defined as those that cannot be produced with conventional techniques (Islam 2014). It turns out that the volume of gas available increases exponentially as conventional gas moves to unconventional gas (Polikar 2011). Figure 1.1b shows that the endowment of conventional gas is estimated at 15,100 Tcf and that of unconventional gas is estimated at 52,700 Tcf except for gas hydrates. Unconventional natural gas has been more difficult and costly to exploit than conventional deposits, until recently. Among these

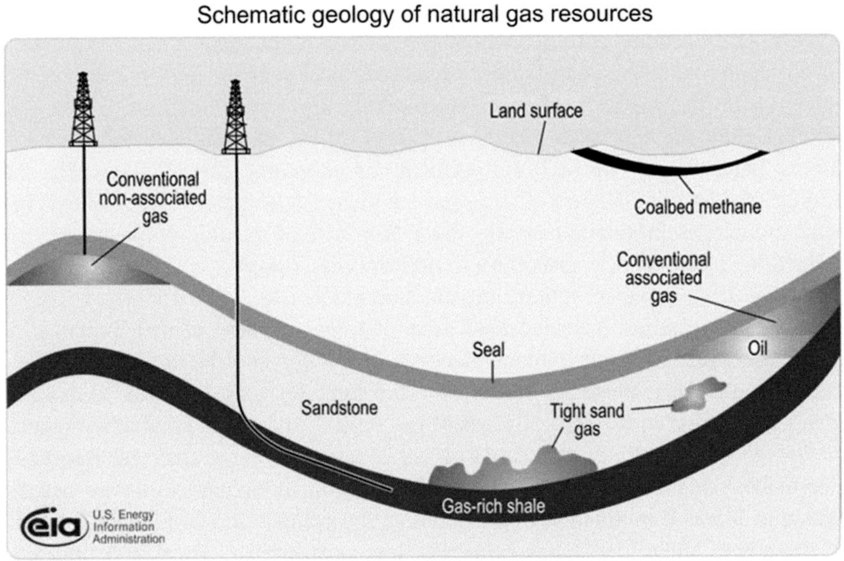

Schematic geology of natural gas resources

Fig. 1.2 Schematic geology of natural gas resources (EIA 2010)

unconventional resources, shale and tight gas exploitation are widely commercialized, with constant improvement of fracturing techniques to increase yield and decrease costs.

Figure 1.2 shows the geologic nature of most major sources of natural gas in schematic form (EIA 2010). Gas-rich shale is the source rock for many natural gas resources, but, until now, has not been a focus for production. Horizontal drilling and hydraulic fracturing have made shale gas an economically viable alternative to conventional gas resources. Conventional gas accumulations occur when gas migrates from gas rich shale into an overlying sandstone formation, and then becomes trapped by an overlying impermeable formation, called the seal. Associated gas accumulates in conjunction with oil, while non-associated gas does not accumulate with oil. Tight sand gas accumulations occur in a variety of geologic settings where gas migrates from a source rock into a sandstone formation, but is limited in its ability to migrate upward due to reduced permeability in the sandstone. Coal bed methane does not migrate from shale, but is generated during the transformation of organic material to coal.

The recent boom in natural gas production in the United States, which has been brought through technical innovations in the recovery of natural gas from previously inaccessible shale rock formations, has helped lower electricity costs and benefitted the petrochemical and manufacturing industries. Even more significantly, it has contributed to a drop in United States carbon dioxide emissions. EIA report (2013b) shows that energy-related carbon dioxide emissions are at their lowest level since 1994 and have fallen 12 % between 2007 and 2012. As a result, inexpensive natural gas accelerates the closure of aging coal plants around the country.

Shale gas formations are usually mature petroleum source rocks where high levels of heat and pressure have converted the source rock material to natural gas. Characteristics of shale gas reservoirs are different from those of typical conventional reservoirs. Shale is a fissile mudstone consisting of silt, $4 \sim 60$ μm, and clay-size particles, less than 4 μm, which are largely mineral fragments (Rezaee 2015). Shale is characterized by thin, parallel, horizontal layers which are formed as cumulative deposits of sedimentary rock (sand, silt, mud, decaying plants and animals and other microorganisms) compressed over long periods of time (millions of years), a process known as compaction. Shale hydrocarbon reservoirs, in addition to mineral fragments, include a small amount of organic matter. Organic material is transformed into hydrocarbon under large overburden stress and high temperature conditions. It also creates a large internal hydrostatic pressure locally, which could cause creation of micro-fracture pores because of the fluid expansion force. The pore size in shale could be less than 2 nm or as high as 2 μm. Nanopores create large capillary pressures, lower the critical pressure, and temperature of hydrocarbon components creating a shift in the phase envelope of the resident fluids, and cause capillary condensation and slippage of gas molecules at the pore walls (Knudsen flow). Because of low matrix permeability, Darcy flow (advection) becomes so small that molecular diffusion can play a significant role in the mass transfer of fluids from the matrix to micro and macro fractures. Hydrocarbon-rich shale reservoirs are typically oil wet while their counterparts, tight sandstones, are generally water wet. In shale gas reservoirs, both free gas and adsorbed gas adsorbed exist. Free gas exists in pore spaces of the matrix and natural fractures, and adsorbed gas is stored on the surface of matrix particles and the faces of natural fractures (Song et al. 2011). Several studies presented that gas desorption may contribute $5 \sim 30$ % of total gas production, but this effects are observed at the late time of well production (Cipolla et al. 2010; Thompson et al. 2011; Mengal and Wattenbarger 2011).

Shale reservoirs have very low permeability and porosity. A typical shale reservoir has a very low permeability matrix of about 1 to 100 nd and a porosity of less than 10 %. To exploit ultra-low permeability reservoirs, hydraulic fracturing technology has been proven to be an effective means. Hydraulic fracturing is a process used in nine out of 10 natural gas wells in the United States, where millions of gallons of water, sand and chemicals are pumped underground to break apart the rock and release the gas (Propublica 2012). Figure 1.3 show the process of hydraulic fracturing. Hydraulic fracturing induces fractures with enormously high permeability and makes fracture networks of interconnected fractures around the wellbore.

Since shale gas reservoirs are relatively thin and infinite laterally, horizontal wells are usually applied to improve production by increasing the contact area of wellbore and the pay zone. The great increase of the surface area of the wellbore facilitates that fluids flow freely from the reservoir to the wellbore. To effectively access the reservoir pores, drilling engineers drill long horizontal wells in the formation parallel to the minimum horizontal stress direction. Then, completion engineers place a large set of multistage transverse hydraulic fractures in each well

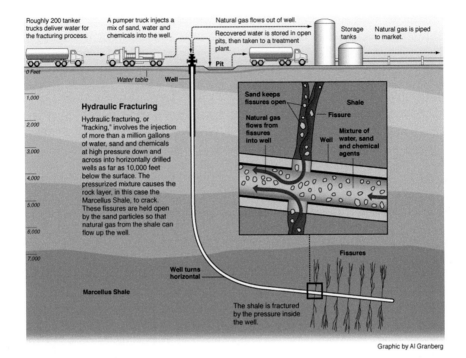

Fig. 1.3 Process of hydraulic fracturing (Propublica 2012)

to stimulate the drainage volume of the well. The horizontal well segment is in the range of 4000–10,000 ft in length (~5000 ft in Eagle Ford, U.S.A. and 9000 ft in Bakken, U.S.A.), consisting of 20–50 transverse hydraulic fractures in the multi-stage stimulation process. Each horizontal well is usually from 350 to 1200 ft apart (350 and 700 ft in Eagle Ford and 1200 ft in Bakken). Flow rates in extremely low permeability shale gas reservoirs depend on the total area of permeable fractures that are hydraulically connected to the well and the matrix permeability of the shale formation. The multistage hydraulic fractures create a dual-porosity environment in the wellbore drainage area, called the "stimulated reservoir volume (SRV)." The dual-porosity environment makes it easier for hydrocarbons to flow from small pores of the matrix, to micro and macro fractures, and to the wellbore. The inverse of this flow hierarchy is much less effective in fluid injection processes. To confirm the dual-porosity nature of the SRV, reservoir engineers compare the permeability from the rate transient test with that of the cores. If the transient-test permeability is much larger than the core permeability, it can be concluded that the hydraulic fracturing process has induced macro-fractures, which, in turn, has created a larger formation effective permeability than that of the matrix.

The central geological properties of a shale gas play are generally assessed in terms of organic geochemistry, organic richness, thickness, thermal maturity, and mineralogy. For the successful production in shale gas reservoirs, high total organic carbon (TOC) content and thermal maturity, relevant thickness, and a low clay content/high brittle mineral content are needed (Rezaee 2015). Shale gas organic geochemistry is a function of the depositional environment and is similar to conventional source rock geochemistry. Lacustrine shale, marine shale, and terrestrial/coal bed shale is typically associated with Type I, II, and III kerogens (Gluyas and Swarbrick 2009; Caineng et al. 2010). Target TOC (wt% kerogen) values are somewhat interrelated to the thickness and other factors that influence gas yield. For commercial shale gas production, Rezaee (2015) notes a target TOC of a least 3 %, while Lu et al. (2012) states that a TOC of 2 % is generally regarded as the lower limit of commercial production in the United States. That said, TOC varies considerably throughout any one shale gas play. The thickness of economic gas shale is one of many considerations. As an example, in North America, the effective thicknesses of shale gas pay zones range from 6 m (Fayetteville, U.S.A.) to 304 m (Marcellus, U.S.A.) (Caineng et al. 2010). Caineng et al. (2010) note a guidance thickness for economic plays of 30guidance thickness for economic plays50 m, where development is continuous and the TOC (wt%) is greater than 2 %. TOC is only an indication of shale gas potential. The actual accumulation of gas from the organic compounds within the shale requires the organic matter to first generate the gas and this is function of the thermal maturity (Lu et al. 2012). Significant shale gas is typically only generated beyond vitrinite reflectance (Ro%) values of approximately 0.7 % (Type III kerogen) to 1.1 % (Type I and II kerogen), which corresponds to depth of between 3.5 and 4.2 km (Gluyas and Swarbrick 2009). However, the most favorable situation is when virtinite reflectance values range from 1.1 to 1.4 (Rezaee 2015). Mineralogy also plays a central role when evaluating gas shale, due to its impact on the performance of fracture treatment. In terms of mineralogy, brittle minerals such as quartz, feldspar, calcite, and dolomite are favorable for the development of extensive fractures throughout the formation in response to fracture treatment. According to Caineng et al. (2010), the brittle mineral content should be greater than 40 % to enable sufficient fracture propagation. Alternatively, Lu et al. (2012) note that in the main shale gas producing areas of the United States, the brittle mineral content is generally greater than 50 % and the clay content is less than 50 %. In more simplistic terms, high clay content results in a more ductile response to hydraulic fracturing, with the shale deforming instead of shattering.

According to EIA report (2014), by 2035, natural gas surpasses coal as the largest source of United States electricity generation. The report anticipated that the share of electricity generated from natural gas grows steadily so that natural gas plants account for more than 70 % of all new capacity. In this situation, shale gas provides the largest source of growth in United States natural gas supply. Figure 1.4 shows the history and prediction of United States natural gas production by source (EIA 2014). The 56 % increase in total natural gas production from 2012 to 2040 in the reference case results from increased development of shale gas, tight gas, and offshore natural gas resources. Shale gas production is the largest contributor, growing by more than 10 Tcf, from 9.7 Tcf in 2012 to 19.8 Tcf in 2040. The shale

Fig. 1.4 United States
natural gas production by
source in the reference case,
1990–2040 (trillion cubic
feet) (EIA 2014)

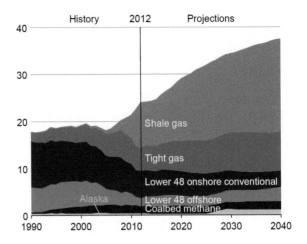

gas share of total United States natural gas production increases from 40 % in 2012
to 53 % in 2040. Tight gas production also increases by 73 % from 2012 to 2040.

Figure 1.5 shows the location of shale basins and the regions analyzed (EIA
2013a). Red colored areas represent the location of basins with shale formations for
which estimates of the risked oil and natural gas in-place and technically recoverable resources were provided. Tan colored areas represent the location of basins
that were reviewed, but for which shale resource estimates were not provided,
mainly due to the lack of data necessary to conduct the assessment. White colored
areas were not assessed in the report. Figure 1.5 shows that there are 137 shale
formations in 41 countries. Estimates of EIA report also provide technically

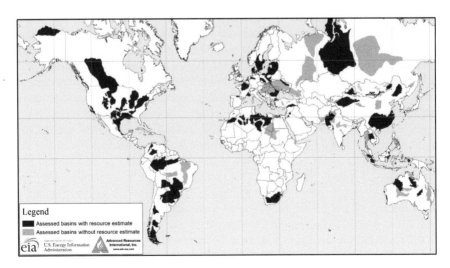

Fig. 1.5 Map of basins with assessed shale oil and shale gas formations (EIA 2013a)

recoverable resources of 345 billion barrels of world shale oil resources and 7299 trillion cubic feet of world shale gas resources (EIA 2013a). The estimates of unproved technically recoverable shale oil and gas resources in country-level detail are presented in Table 1.1 (EIA 2013a).

Table 1.1 Unproved technically recoverable shale gas and oil resources in total world (million barrels) (EIA 2013a)

Region totals and selected countries	2013 EIA/ARI unproved wet shale gas technically recoverable resources (TRR)	2013 EIA/ARI unproved shale oil technically recoverable resources (TRR)
Europe	**470**	**12,900**
Bulgaria	17	200
Denmark	32	0
France	137	4700
Germany	17	700
Netherlands	26	2900
Norway	0	0
Poland	148	3300
Romania	51	300
Spain	8	100
Sweden	10	0
United Kingdom	26	700
Former Soviet Union	**415**	**77,200**
Lithuania	0	300
Russia	287	75,800
Ukraine	128	1100
North America	**1685**	**80,000**
Canada	573	8800
Mexico	545	13,100
United States	567	58,100
Asia and Pacific	**1607**	**61,000**
Australia	437	17,500
China	1115	32,200
Indonesia	46	7900
Mongolia	4	3400
Thailand	5	0
South Asia	**201**	**12,900**
India	96	3800
Pakistan	105	9100
Middle East and North Africa	**1003**	**42,900**

(continued)

Table 1.1 (continued)

Region totals and selected countries	2013 EIA/ARI unproved wet shale gas technically recoverable resources (TRR)	2013 EIA/ARI unproved shale oil technically recoverable resources (TRR)
Algeria	707	5700
Egypt	100	4600
Jordan	7	100
Libya	122	26,100
Morocco	12	0
Tunisia	23	1500
Turkey	24	4700
Western Sahara	8	200
Sub-Saharan Africa	**390**	**100**
Mauritania	0	100
South Africa	390	0
South America and Caribbean	**1430**	**59,700**
Argentina	802	27,000
Bolivia	36	600
Brazil	245	5300
Chile	48	2300
Colombia	55	6800
Paraguay	75	3700
Uruguay	2	600
Venezuela	167	13,400
Total world	**7201**	**345,000**

Note The bold values indicate total TRR of each continent

The possibility of cheaper and cleaner energy from shale gas has prompted interest from governments around the world. If it can achieve the necessary innovations for tapping perhaps the largest shale gas reserves in the world, China may be able to reduce its dependence on coal and shift to a lower-carbon economy (Tian et al. 2014). European countries such as the United Kingdom are also exploring the possibility of exploiting shale gas

However, caution is warranted. The large deployment of fracking technology faces significant hurdles outside of the United States context. China's nascent industry is plagued by technical bottlenecks, lack of adequate water supply, and poor infrastructure (Hu and Xu 2013). Drilling an exploratory shale gas well in China still costs much more than it does in the United States. In Europe, the challenges are more likely to be political and legal (Helm 2012). Unlike in the United States, European landowners do not automatically own the rights to extract the resources from the ground beneath their property, making the building of new extraction plants fraught with political difficulties (Gold 2014).

Shale gas therefore has the potential to be very significant source of natural gas, and has the potential to greatly increase the gas resource of many nations across the globe. As outlined by Ridley (2011), the significance and future of shale gas will be influenced by the interplay of a wide variety of other issues, including the following: potentially falling gas prices due to increased production, increased demand for gas due to increased adoption of natural gas to produce energy, and reduced production costs due to technological development. Among these, for the understanding of shale gas reservoir, this book provides comprehensive technical information in terms of petroleum reservoir engineering.

1.3   Historical Review

The first use of shale gas in the US can be traced back to 1821, when a shallow well drilled in the Devonian Dunkirk Shale in Chautauqua County, New York (Table 1.2). The natural gas was produced, transported and sold to local establishments in the town of Fredonia (Peebles 1980; David et al. 2004). Following this discovery, hundreds of shallow shale wells were drilled along the Lake Erie shoreline and eventually several shale gas fields were established southeastward from the lake in the late 1800 s (David et al. 2004). However, shale gas production had been discouraged because much larger volumes natural gas could produce from conventional reservoirs as with the Drake Well developed in 1859 (Table 1.2) (Peebles 1980). These main stages in the shale gas industry from 1860 to 1970s were shale gas reservoirs discovered in the western Kentucky in 1863, in West Virginia in the 1920s, and hydraulic fracturing first used in the 1940s (Table 1.2).

Table 1.2 Selected progress of the shale gas development in the US between 1821 and 1940s (Wang et al. 2014)

Time	Brief introduction
1821	In 1821, the first well was drilled in the Devonian Dunkirk Shale in Chautauqua County, New York. The natural gas was used to illuminate the town of Fredonia
1859	The Drake Well was developed in 1859 at Cherry tree Township, Venango County in the northwestern Pennsylvania. The Drake Well demonstrates that oil can be produced in large volumes. Hence, the Drake Well is viewed as one of the most important oil well ever drilled
1860s–1930s	(i) Shale-gas development spread westward along the southern shore of Lake Erie and reached northeastern Ohio in the 1870s. In 1863, gas was discovered in the western Kentucky part of the Illinois basin (ii) By the 1920s, drilling for shale gas had progressed into West Virginia, Kentucky, and Indiana (iii) By 1926, the Devonian shale gas fields of eastern Kentucky and West Virginia comprised the largest known gas occurrences in the world
Late 1940s	Hydraulic fracturing first used to stimulate oil and gas wells. The first hydraulic fracturing treatment was pumped in 1947 on a gas well operated by Pan American Petroleum Corporation in Grant County, Kansas

The 1973 and 1979 oil crises had led the U.S. to address energy shortages, and high price of oil. The oil crisis in 1970s propelled the U.S. government to invest in research and development and demonstration of alternative energy, including natural gas from shale formations. Meanwhile, the high oil prices attracted private enterprises to invest in unconventional natural gas (Henriques and Sadorsky 2008; Cleveland 2005; Bowker 2007; Montgomery 2005).

Before 1970s, deep shale gas, such as the Barnett Shale in Texas and Marcellus in Pennsylvania, has been known but believed to have extremely low permeability and thus were not considered economically feasible (Curtis 2002; NETL 2011; Loucks and Ruppel 2007). In the late 1970s, the U.S. Department of Energy (DOE) initiated the Eastern Gas Shale Project (EGSP) as a series of geological, geochemical, and petroleum engineering studies to evaluate the gas potential and to enhance gas production from the extensive Devonian and Mississippian organic-rich black shale within the Appalachian, Illinois, and Michigan basins in the eastern U.S. (Soeder 1988; Curtis 2002; NETL 2011; Loucks and Ruppel 2007). In addition to providing R&D support, the Gas Research Institute (GRI) was established in 1977 (AGF 2007). The GRI was providing central organizations to manage the public research programs that were funded via mechanisms designed to pass research and development (R&D) costs through to the end-customer. A few years later, the DOE was established and funding for energy R&D, in general, and in particular, supplemental gas supplies, were substantially increased. During the 1980s and early 1990s, GRI was expanded to include R&D programs addressing supply, transmission, distribution and end-use. In the late 1990s, the National Energy Technology Laboratory (NETL) was established. A consolidated research program led by NETL was initiated aimed primarily at preventing pipeline damage of the aging natural gas infrastructure in the U.S. In the same time period, GRI was reorganized to emphasize near-term industry impact. In 2000, GRI and the Institute of Gas Technology (IGT), which had been the R&D performing laboratory for the gas distribution industry, merged to form the Gas Technology Institute (GTI) (Fig. 1.6) (AGF 2007).

Meanwhile, some pioneering oil and gas companies had tried to combine larger fracture designs, rigorous reservoir characterization, horizontal drilling, and lower cost approaches to hydraulic fracturing to make the extraction shale gas economic (EIA 2011; Montgomery 2005; Kvenvolden 1993). The best-known pioneering company is the Mitchell Energy & Development Corp. The company went on to test various processes of hydraulic fracturing to exploit natural gas in the Barnett Shale formation in North Texas between 1981 and the early 1990s. Production from many of the 30 or so test wells fell short of covering operational costs. The company focused on the test results yielding the greatest returns. The engineers of this company analyzed and retested until eventually, the successful use of hydraulic fracturing to drill into shale formation for natural gas was completed (EIA 2011; Gidley 1989; Fjær et al. 2008; Kutchin 2001; Gardner and Canning 1994). The hydraulic fracturing techniques developed by the Mitchell Energy & Development Corp. changed the face of the oil and gas industry (EIA 2011; Pickett 2010; Gidley 1989; Kutchin 2001; Becchetti et al. 2005).

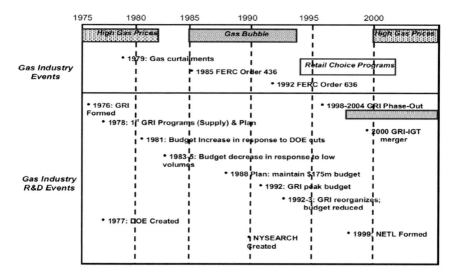

Fig. 1.6 A timeline of United States official gas industry research institution (AGF 2007)

In a word, these efforts from government and private enterprise during this period contributed to the rapid growth in output of shale gas. The output of shale gas in the U.S. increased more than seven-fold between 1979 and 2000 (EIA 1999).

Since 2000, three factors have contributed to increase energy companies' confidence in the ability to profitably produce natural gas from the shale formation. Above all, the drilling techniques are more advanced. In 2002, Devon Energy Corp. invested $3.5 billion in cash and stock to acquire Mitchell Energy & Development Corp. Devon Energy Corp added horizontal drilling to its repertoire to make shale gas wells even more productive. In the few short years since then, technology has continued to improve: drilling techniques have continued to advance, and horizontal drilling has been employed by many exploration and production companies in search of unconventional resources. The use of horizontal drilling in conjunction with hydraulic fracturing greatly expanded the ability of producers to profitably produce natural gas from low permeability shale formations (EIA 2011; Wang 2011; Pickett 2010; Bowker 2007; Kutchin 2001; Martineau 2007; Wang and Chen 2012).

In addition, the rise in oil and gas prices since 2003 made shale gas more economically attractive than ever before (Owen et al. 2010; Heinberg and Fridley 2010). From the mid-1980s to 2003, the price of crude oil was generally under $25/barrel (BP 2011). The crude oil price rose above $30/barrel in 2003, reached $60/barrel in 2005, exceed $75/barrel in 2006, reached nearly $100/barrel in 2007, and peaked over $140/barrel in 2008. Finally, the prospect of falling conventional gas production of U.S. since 2000 triggered expectations of higher gas price inflation in. As shown in Fig. 1.7, U.S. gas production was in slow but steady decline in the early 2000s. In the early 2000s, it was expected that U.S. natural gas

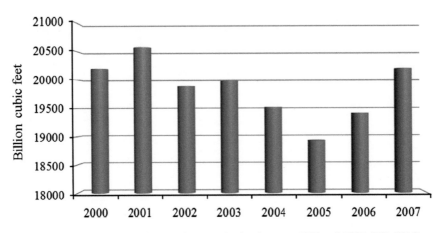

Fig. 1.7 United States domestic natural gas production between 2000 and 2007 (EIA 2015)

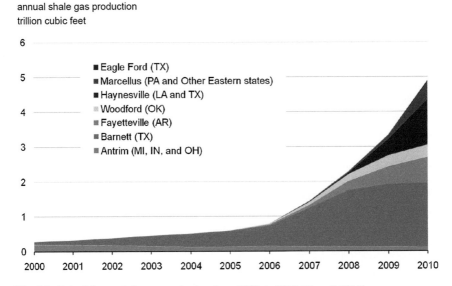

Fig. 1.8 United States shale gas production from 2000 to 2010 (Newell 2011)

price would rise in response to the resulting tight market (Burke 2012; Rogers 2011; The Perryman Group 2007).

Due to growing confidence in their ability to profitably produce natural gas in shale formations, the upstream oil and gas companies aggressively entered the shale gas business. Drilling for gas has increased sharply by the independent energy companies such as Devon Energy, Goodrich Petroleum, and XTO Energy. This can be shown by the development of the Barnett Shale Play, the largest producible reserves of any onshore natural gas field in the U.S. at that time (Fig. 1.8)

Fig. 1.9 A comparison of the numbers of shale gas well in 1997 and in 2009 in the Barnett Shale Play (Newell 2011)

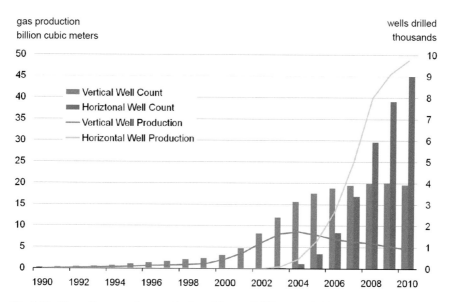

Fig. 1.10 The well count and gas production of Barnett field between 1990 and 2010. (Newell 2011)

(Jarvie et al. 2007; Bowker 2007; The Perryman Group 2007). From 1997 to 2009, more than 13,500 gas wells have been drilled in the Barnett Shale Play (Fig. 1.9). Naturally, the output of natural gas from the Barnett Shale Play increased sharply (Fig. 1.10). In 2004, gas production from the Barnett Shale Play overtook the level of shallow shale gas production from historic shale plays such as the Appalachian Ohio Shale and Michigan Basin Antrim plays (NETL 2011).

Inspired by the success of Barnett Shale Play, oil and gas companies rapidly entered other shale formation, including the Fayetteville Haynesville, Marcellus, Woodford, Eagle Ford and other shale plays (EIA 2011). The proliferation of activity in these new plays has increased shale gas production in the U.S. from 1.0 trillion cubic feet in 2006 to 4.87 trillion cubic feet, or 23 % of total U.S. natural gas production in 2010 (EIA 2011).

1.4 Scope and Organization

Although shale gas industry has made rapid progress in recent years, there are great gaps of knowledge within industry. Because shale gas reservoirs have distinctive features dissimilar to those of conventional reservoirs, an accurate evaluation on the behavior of shale gas reservoirs needs an integrated understanding on the characteristics of reservoir and fluids. This book fills the need for integrative approach necessary to understand the shale gas reservoir. It covers general overview of shale gas reservoirs such as natural fracture system, adsorption/desorption of methane, diffusion in nanopores, and non-linear flow in the reservoir. In addition, the subject of geomechanical modeling, which is of importance in ultra-low permeability reservoir, is presented in detail. Based on the proposed model, pressure transient and production characteristics of a fractured horizontal well in shale gas reservoir are analyzed with respect to reservoir and fracture properties. Methods for evaluation of properties in shale gas reservoir are also provided. Further, new subject, CO_2 injection and advanced well structure in shale gas reservoir, is contained. This book provides insight into integrative understanding of shale gas reservoir for state-of-the-art characteristic methods of shale formation and long-term production performance.

This book is composed of five Chapters. These include Introduction, Characteristics of Shale Reservoirs, Numerical Modeling, Performance Analysis, and Future Technologies. In the Introduction, the objectives of this book and general features of shale gas reservoir is described. Characteristics of Shale Reservoirs Chapter shows the specific features of shale gas reservoir such as natural fracture system, adsorption of methane, diffusion in nanopores, non-Darcy flow, and stress-dependent compaction in the shale reservoirs. Based on these complex features of shale reservoirs, various aspects of numerical simulation are described in the Numerical Modeling Chapter. In these sections, numerical model is verified with field data by history matching technique. In the Performance Analysis Chapter, methods for evaluation of shale reservoir properties and production behavior are introduced. Mini-frac test, decline curve analysis (DCA), and rate transient test (RTA) are introduced for estimating the initial pressure, permeability, fracture half-length, and so forth. Study for evaluating of pressure behavior in shale gas reservoir is also introduced. Finally, in the Future Technologies Chapter, technique of CO_2 injection in the shale reservoir and advanced well structure are presented. Feasibility of CO_2 injection in the shale is analyzed for the EGR and CO_2 storage. Several studies of advanced well structure including fish bone well are introduced.

References

Aguilera R (2014) Flow units: from conventional to tight-gas to shale-gas to tight-oil to shale-oil reservoirs. SPE Res Eval Eng 17(2):190–208. doi:10.2118/165360-PA

Aguilera R et al (2008) Natural gas production from tight gas formations: a global perspective. Paper presented at the 19th world petroleum congress, Madrid, Spain, 29 June–3 July 2008

American Gas Foundation (2007) Research and development in natural gas transmission and distribution. Washington D.C

Becchetti L et al (2005) Corporate social responsibility and corporate performance: evidence from a panel of US listed companies. CEIS Tor Vergata, Research paper series 26(78)

Bowker KA (2007) Barnett shale gas production, Fort Worth basin: issues and discussion. AAPG Bull 91:523–533

BP (2011) BP statistical review of world energy June 2011, London

Burke D (2012) Exxon's big bet on shale gas. http://fortune.com/2012/04/16/exxons-big-bet-on-shale-gas. Accessed 16 Apr 2012

Caineng et al (2010) Geological characteristics and resource potential of shale gas in China. Petrol Explor Dev 37(6):641–653

Cipolla CL et al (2010) Reservoir modeling in shale-gas reservoirs. SPE Res Eval Eng 13 (4):638–653. doi:10.2118/125530-PA

Cleveland CJ (2005) Net energy from the extraction of oil and gas in the United States. Energy 30 (5):769–782

Curtis JB (2002) Fractured shale-gas systems. AAPG Bull 86(11):1921–1938

David G et al (2004) Fractured shale gas potential in New York. Northeast Geol Env Sci 26:57–78

EIA (1999) U.S. crude oil, natural gas, and natural gas liquids reserves 1998 annual report. Office of oil and gas. U.S. Department of Energy, Washington, D.C

EIA (2010) Schematic geology of natural gas resources. http://www.eia.gov/oil_gas/natural_gas/special/ngresources/ngresources.html. Accessed 27 Jan 2010

EIA (2011) World shale gas resources: an initial assessment of 14 regions outside the United States. Office of energy analysis. U.S. Department of Energy, Washington, D.C

EIA (2013a) Technically recoverable shale oil and shale gas resources: an assessment of 137 shale formations in 41 countries outside the United States. Independent statistics and analysis. U.S. Department of Energy, Washington, D.C

EIA (2013b) US energy related carbon dioxide emissions 2012. Independent statistics and analysis. U.S. Department of Energy, Washington, D.C

EIA (2014) Annual energy outlook 2014 with projections to 2040. Office of integrated and international energy analysis. U.S. Department of Energy, Washington, D.C

EIA (2015a) Annual energy outlook 2015 with projections to 2040. Office of integrated and international energy analysis. U.S. Department of Energy, Washington, D.C

EIA (2015b) U.S. natural gas marketed production. https://www.eia.gov/dnav/ng/hist/n9050us2a.htm. Accessed 30 Nov 2015

Fjær E et al (2008) Petroleum related rock mechanics. Elsevier, Netherlands

Gardner GHF, Canning AHF (1994) Effects of irregular sampling on 3-D prestack migration. Paper presented at the annual meeting of the society of exploration geophysicists and international exposition. Los Angeles, California 23–27 Oct 1994

Gidley JL (1989) Recent advances in hydraulic fracturing. Society of Petroleum Engineers, Texas, Richardson

Gluyas J, Swarbrick R (2009) Petroleum geoscience. Blackwell, Malden, Massachusetts

Gold R (2014) The boom: how fracking ignited the American energy revolution and changed the world. Simon and Schuster, New York

Heinberg R, Fridley D (2010) The end of cheap coal. Nature 468:367–369

Helm D (2012) The carbon crunch: how we're getting climate change wrong-and how to fix it. Yale University Press, New Haven, Connecticut

Henriques I, Sadorsky P (2008) Oil prices and the stock prices of alternative energy companies. Energy Econ 30(3):998–1010

Hu D, Xu S (2013) Opportunity, challenges and policy choices for China on the development of shale gas. Energy Policy 60:21–26

Islam MR (2014) Unconventional gas reservoirs evaluation, appraisal, and development. Gulf Professional Publishing, Houston, Texas

Jarvie DM et al (2007) Unconventional shale-gas systems: the Mississippian Barnett shale of north-central Texas as one model for thermogenic shale-gas assessment. AAPG Bull 91 (4):475–499

Kutchin JW (2001) How Mitchell Energy & Development Corp. Got its start and how it grew: an oral history and narrative overview. Universal Publishers, Boca Raton, Florida

Kvenvolden KA (1993) Gas hydrates-geological perspective and global change. Rev Geophys 31 (2):173–187

Loucks RG, Ruppel SC (2007) Mississippian Barnett shale: lithofacies and depositional setting of a deep-water shale-gas succession in the Fort Worth basin, Texas. AAPG Bull 91(4):579–601

Lu S et al (2012) Classification and evaluation criteria of shale oil and gas resources: discussion and application. Pet Explor Dev 39(2):268–276

Martineau DF (2007) History of the Newark East field and the Barnett shale as a gas reservoir. AAPG Bull 91(4):399–403

Mengal SA, Wattenbarger RA (2011) Accounting for adsorbed gas in shale gas reservoirs. Paper presented at the SPE Middle East oil and gas show and conference. Manama, Bahrain, 25–26 Sept 2011

Montgomery SL (2005) Mississippian Barnett shale, Fort Worth basin, north-central Texas: gas-shale play with multi-trillion cubic foot potential. AAPG Bull 89(2):155–175

Newell R (2011) Shale gas and the outlook for U.S. natural gas markets and global gas resources. EIA. http://www.eia.gov/pressroom/presentations/newell_06212011.pdf

NETL (2011) Shale gas: applying technology to solve America's energy challenges. U.S. Department of Energy, Washington, D.C

Owen NA et al (2010) The status of conventional world oil reserves-hype or cause for concern? Energy Policy 38(8):4743–4749

Peebles MWH (1980) Evolution of the gas industry. New York University Press, New York

Pickett A (2010) Technologies, methods reflect industry quest to reduce drilling footprint. The American Oil and Gas Reporter, Consumer Energy Alliance, Houston, Texas

Polikar M (2011) Technology focus: unconventional resources. J Pet Tech 63(7):98. doi:10.2118/ 0711-0098-JPT

Propublica (2012) What is hydraulic fracturing? http://www.propublica.org/special/hydraulic-fracturing-national. Accessed 7 March 2012

Rezaee R (ed) (2015) Fundamental of gas shale reservoirs. Wiley, New Jersey

Ridley M (2011) The shale gas shock, report 2. The Global Warming Policy Foundation, London, UK

Rogers H (2011) Shale gas-the unfolding story. Oxf Rev Econ Policy 27(1):117–143

Soeder DJ (1988) Porosity and permeability of Eastern Devonian gas shale. SPE Form Eval 3(1):116–124

Song B et al (2011) Design of multiple transverse fracture horizontal wells in shale gas reservoirs. Paper presented at the SPE hydraulic fracturing technology conference. Woodlands, Texas, 24–26 Jan 2011

The Perryman Group (2007) Bounty from below: the impact of developing natural gas resources associated with the Barnett shale on business activity in Fort Worth and the surrounding 14-county area. Waco, Texas

Thompson JM et al (2011) Advancements in shale gas production forecasting-a Marcellus case study. Paper presented at the SPE Americas unconventional gas conference and exhibition, Woodlands, Texas, 14–16 June 2011

Tian L et al (2014) Stimulating shale gas development in China: a comparison with the U.S. experience. Energy Policy 75:109–116

Wang Q (2011) Time for commercializing non-food biofuel in China. Renew Sust Energ Rev 15(1):621–629

Wang Q, Chen X (2012) China's electricity market-oriented reform: from an absolute to a relative monopoly. Energy Policy 51:143–148

Wang Q et al (2014) Natural gas from shale formation—the evolution, evidences and challenges of shale gas revolution in United States. Renew Sust Energ Rev 30:1–28

Chapter 2
Characteristics of Shale Reservoirs

2.1 Introduction

Shale gas reservoir shows several features dissimilar with conventional reservoir which make it difficult to understand behavior of it. In this chapter, these features such as natural fracture system, adsorption/desorption of gas, diffusion in nanopores, non-Darcy flow, and stress-dependent compaction are presented. In general, shale gas reservoir includes natural fractures which are believed to play a significant role in hydraulic fracture propagation and gas production. Pressure behavior of dual porosity model used to simulate the natural fracture system is presented. In shale reservoirs, hydrocarbon gas is stored in two ways which are free gas in the pore media and absorbed gas in the surface of organic material. Previous studies presented that gas desorption contribute 5–30 % of total gas production in shale reservoir. In order to simulate gas production in shale gas reservoirs, an accurate model of gas adsorption is very important. According to the International Union of Pure and Applied Chemistry (IUPAC) standard classification system, there are six different types of adsorption. Among these, Langmuir isotherm and BET isotherm are suitable for shale formation so that they are presented. Due to the presence of nanopores, fluid flow in shale reservoir cannot be calculated from Darcy equation. This phenomenon can be explained by the concept of slip flow and a common way to model the flow of gas in nanopore is to modify the no-slip boundary condition in continuum models by accounting for a slip boundary condition. Several studies are suggested for accurate diffusion modeling of shale reservoir. Darcy equation is also cannot applied to hydraulic fractures due to high velocity of gas flow. When the gas velocity increases significant inertial (non-Darcy) effects can occur. This induces an additional pressure drop in the hydraulic fractures in order to maintain the production rate. For simulate this mechanism, Forchheimer equation which can replace

© The Author(s) 2016
K.S. Lee and T.H. Kim, *Integrative Understanding of Shale Gas Reservoirs*,
SpringerBriefs in Applied Sciences and Technology,
DOI 10.1007/978-3-319-29296-0_2

the Darcy equation is used. The conductivity of fracture network in shale reservoir is sensitive to the stress-dependent compaction effect. Change of porosity and permeability due to change of stress and strain should be considered. Effect of shale rock compaction can be considered by several stress-dependent correlations coupled with geomechanical model.

2.2 Natural Fracture System

A naturally fractured reservoir has been referred to as dual porosity system because two types of porous regions that present distinctly different properties are in presence (Barenblatt et al. 1960). The first region forms the continuous system connected with the wells, whereas the second region only feeds fluid locally to the first region. These regions represent matrix and fractures which have different fluid storage and conductivity characteristics in shale gas reservoirs.

Warren and Root (1963), who idealized the system as an orthogonal set of intersecting fractures and cubic matrix blocks (Fig. 2.1), invoked a simple pseudo-semi-steady-state (PSSS) model of transfer from matrix to fractures. Figure 2.2 shows the early part of fundamental pressure response of Warren and Root (1963)'s dual porosity model in semi-log plot (Stewart 2011). Pressure behavior is characterized by the first straight line, a transition which looks like a straight line of nearly a zero slope, and a final straight line displaying the same slope as the first. The first straight line usually shows very short duration and represents the fracture system alone. The equations of this line are

Fig. 2.1 Naturally fracture reservoir model composed of an orthogonal set of intersecting fractures and cubic matrix blocks

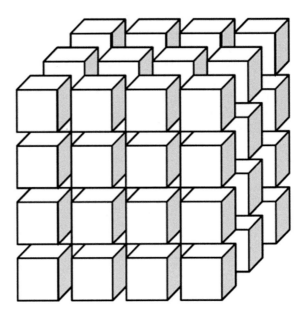

Fig. 2.2 Dual porosity
construction on a semi-log
graph (Stewart 2011)

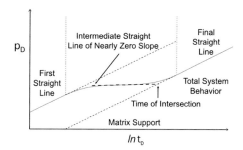

$$p_{wD} = \frac{1}{2}(\ln t_D) - \ln \omega + \ln \frac{4}{\gamma} \quad \text{or} \tag{2.1}$$

$$p_{wf} = p_i - \frac{q_{sc}B\mu}{4\pi k_{fb}h}\left(\ln t + \ln \frac{4k_{fb}}{\phi_{fb}c_f\mu r_w^2\gamma}\right) \tag{2.2}$$

where p_{wD} is the dimensionless well flowing pressure, t_D the dimensionless time, ω the dimensionless storativity, γ the exponential of Euler's constant, 1.781 or $e^{0.5772}$, p_{wf} the well flowing pressure p_i the initial reservoir pressure q_{sc} the rate at standard condition, B the formation volume factor, μ the viscosity, k_{fb} the bulk fracture permeability, h the net pay thickness, t the time, ϕ_{fb} the bulk fracture porosity, c_f the formation compressibility, and r_w the well radius. The final straight line represents total system behavior and the equations of this period are given by

$$p_{wD} = \frac{1}{2}\left(\ln t_D + \ln \frac{4}{\gamma}\right) \quad \text{or} \tag{2.3}$$

$$p_{wf} = p_i - \frac{q_{sc}B\mu}{4\pi k_{fb}h}\left[\ln t + \ln \frac{4k_{fb}}{(\phi c_t)_{m+f}\mu r_w^2\gamma}\right] \tag{2.4}$$

where c_t is the total compressibility. Subscripts m and f indicate matrix and natural fracture. The separation between the lines is $\ln \omega$, which becomes larger in absolute value as ω become smaller. In Fig. 2.3, fracture and matrix pressures which present a good insight into the mechanism of dual porosity behavior have been illustrated (Stewart 2011). At very early time, initial fracture and matrix pressures are same, whereupon support flow from the matrix is negligible. As the pressure transient propagates out from the well, fracture pressure declines quickly and matrix pressure declines slowly due to difference of conductivity. The flattening of semi-log graph is due to this period increasing support from the matrix to fracture. Depending on slowing down the rate of change of the fracture pressure and catching up of the matrix pressure, the two pressures are nearly analogous. Total system behavior is reached when the media pressures attain this dynamic equilibrium.

Fig. 2.3 Fracture and matrix
pressure in the natural fracture
system (Stewart 2011)

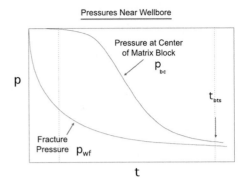

2.3 Adsorption

The organic matter in shale has a strong adsorption potential due to the large surface
area and affinity to methane (Yu et al. 2014). In order to simulate gas production in
shale gas reservoirs, an accurate model of gas adsorption is very important.
According to the standard classification system of the International Union of Pure
and Applied Chemistry (IUPAC) (Sing et al. 1985), there are six different types of
adsorption, as shown in Fig. 2.4. The shape of the adsorption isotherm is closely
related to the properties of adsorbate and solid adsorbent, and on the pore-space
geometry (Silin and Kneafsey 2012). The detailed description of the six isotherm
classifications can be found in Sing et al. (1985).

Fig. 2.4 Types of physical
sorption isotherm (Sing et al.
1985)

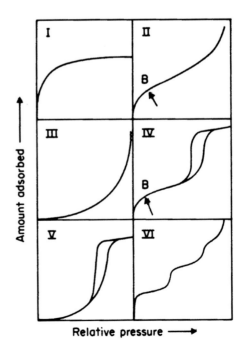

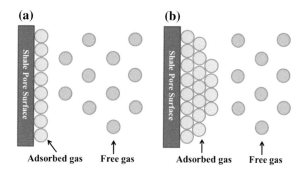

Fig. 2.5 The schematic plots of monolayer and multilayer gas adsorption (Yu et al. 2014). **a** Monolayer Langmuir adsorption **b** Multilayer BET adsorption

The most commonly applied adsorption model for shale gas reservoirs is the classic Langmuir isotherm (Type I) (Langmuir 1918), which is based on the assumption that there is a dynamic equilibrium at constant temperature and pressure between adsorbed and non-adsorbed gas. Also, it is assumed that there is only a single layer of molecules covering the solid surface, as shown in Fig. 2.5a. The Langmuir isotherm has two fitting parameters as shown below:

$$V = \frac{V_L p}{p + p_L},$$
(2.5)

where V is the gas volume of adsorption at pressure p, V_L the Langmuir volume or the maximum gas volume of adsorption at the infinite pressure, and p_L the Langmuir pressure, which is the pressure corresponding to one-half Langmuir volume. Instantaneous equilibrium of the sorbing surface and the storage in the pore space is assumed to be established for the Langmuir isotherm (Freeman et al. 2012). Gao et al. (1994) demonstrated that the instantaneous equilibrium is a reasonable assumption because the ultra-low permeability in shale leads to very low gas flow rate through the kerogen component of shale.

At high reservoir pressures, one can expect that natural gas sorbed on the organic carbon surfaces forms multi-molecular layers. In other words, the Langmuir isotherm may not be a good approximation of the amount of gas sorbed on organic carbon-rich mudrocks. Instead, multilayer sorption of natural gas should be expected on organic carbon surfaces, and the gas adsorption isotherm of Type II should be a better choice. Type II isotherm often occurs in a non-porous or a macroporous material (Kuila and Prasad 2013). Brunauer et al. (1938) suggested the BET isotherm model which is a generalization of the Langmuir model to multiple adsorbed layers, as shown in Fig. 2.5b. The expression is shown as follows:

$$V_L = \frac{V_m C p}{(p_o - p)\left[1 + \frac{(C-1)p}{p_o}\right]}$$
(2.6)

where V_m is the maximum adsorption gas volume when the entire adsorbent surface is being covered with a complete monomolecular layer, C a constant related to the net heat of adsorption, and p_o the saturation pressure of the gas. C is defined as below:

$$C = \exp\left(\frac{E_1 - E_L}{RT}\right), \tag{2.7}$$

where E_1 is the heat of adsorption for the first layer, E_L the heat of adsorption for the second and higher layers and is equal to the heat of liquefaction, R the gas constant, and T the temperature. The assumptions in the BET theory include homogeneous surface, no lateral interaction between molecules, and the uppermost layer is in equilibrium with gas phase. A more convenient form of the BET adsorption isotherm equation is as follows:

$$\frac{p}{V(p_o - p)} = \frac{1}{V_m C} + \frac{C - 1}{V_m C}\frac{p}{p_o} \tag{2.8}$$

A plot of $\frac{p}{V(p_o - p)}$ against $\frac{p}{p_o}$ should give a straight line with intercept of $\frac{1}{V_m C}$ and slope of $\frac{C-1}{V_m C}$. Based on V_m, the specific surface area can be calculated using the following expression:

$$S = \frac{V_m N a}{22,400} \tag{2.9}$$

where S is the specific surface area in m^2/g, N the Avogadro constant (number of molecules in one mole, 6.023×10^{23}), a the effective cross-sectional area of one gas molecule in m^2, and 22,400 is the volume occupied by one mole of the adsorbed gas at standard temperature and pressure.

The standard BET isotherm assumes that the number of adsorption layers is infinite. But, in the case of n adsorption layers in some finite number, then a general form of BET isotherm is given below:

$$V(p) = \frac{V_m C \frac{p}{p_o}}{1 - \frac{p}{p_o}}\left[\frac{1 - (n+1)\left(\frac{p}{p_o}\right)^n + n\left(\frac{p}{p_o}\right)^{n+1}}{1 + (C-1)\frac{p}{p_o} - C\left(\frac{p}{p_o}\right)^{n+1}}\right] \tag{2.10}$$

where n is the maximum number of adsorption layers. When $n = 1$, Eq. 2.10 will be reduced to the Langmuir isotherm, Eq. 2.5. When $n = \infty$, Eq. 2.10 will be reduced to Eq. 2.6.

Figure 2.6 compares shapes of the Langmuir and BET isotherms. Gas desorption along the BET isotherm contributes more significantly at early time of production than that with the Langmuir isotherm curve. This is because the slope of the BET isotherm curve at high pressure is larger than that of the Langmuir isotherm curve,

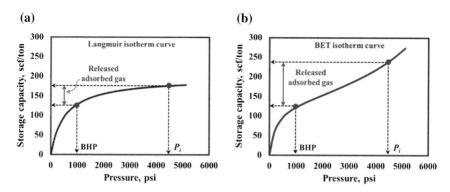

Fig. 2.6 Comparison of the Langmuir and BET isotherms (Yu et al. 2014). **a** Langmuir isotherm
b BET isotherm

resulting in more adsorbed gas releasing at early production times. In addition, under the same pressure drop from the initial reservoir pressure to the bottomhole pressure, the amount of released adsorbed gas with the BET isotherm curve is larger than that with the Langmuir isotherm curve.

2.4   Diffusion

The Darcy equation has been used for more than 150 years to linearly relate fluid-flow rate and pressure gradient across a porous system. The linearity of the Darcy equation makes it easy and practical to use in reservoir engineering analysis and numerical reservoir simulations. However, physics of fluid flow in shale reservoirs cannot be predicted from standard flow or mass transfer models because of the presence of nanopores, ranging in size from one to hundreds of nanometers, in shales. Conventional continuum flow equations, Darcy's law, greatly underestimate the flow rate when applied to nanopore-bearing shale reservoirs.

In other to articulate this phenomenon, Brown et al. (1946) suggested the concept of slip flow, which provided an explanation for the observed relationship between gas flow rate and mean pressure. As mentioned earlier, the pores in producing shale gas reservoirs are in the range of 1–100 nm so that the gas molecules contained in the pores are of comparable size ($\sim$0.5 nm). Under certain pressure and temperature conditions, the distance between gas molecules (mean free path) exceeds the size of the pores. In such conditions, the gas molecules might move singly through the pores and the concept of continuum and bulk flow may not be applicable. Knudsen number, K_n, is the ratio of mean free path, λ, to pore diameter, d, and can be used to identify different flow regimes in the porous media as given below

Table 2.1 Different flow regimes as a function of Knudsen number (Rezaee 2015)

Knudsen number (K_n)	Flow regime
$0-10^{-3}$	Continuum/darcy flow (no-slip flow)
$10^{-3}-10^{-1}$	Slip flow
$10^{-1}-10^{1}$	Transition flow
$10^{1}-\infty$	Free-molecule flow

$$K_n = \frac{\lambda}{d}, \tag{2.11}$$

where

$$\lambda = \frac{k_B T}{\sqrt{2}\pi\delta^2 p}, \tag{2.12}$$

in which k_B is the Boltzmann constant and δ is the collision diameter of the gas molecule. Table 2.1 presents flow regimes corresponding to Knudsen number ranges (Rathakrishnan 2004; Rezaee 2015). Continuum no-slip flow or Darcy equation is valid for $K_n < 10^{-3}$. Continuum flow with slip correction (Klinkenberg) is valid for $K_n < 10^{-1}$, which covers most conventional gas reservoirs and many tight gas reservoir conditions as well. However, as a result of the existence of nanopores in shales, the molecular mean free path becomes comparable with the characteristic geometric scale and K_n could be larger than 0.1. Under this condition, Knudsen diffusion, in addition to correction for the slip boundary condition, becomes the dominant mechanism and hence new forms of gas flow equations are needed. Various gas flow models for slip flow and Knudsen diffusion will be introduced in this section.

Klinkenberg (1941) showed experimentally that a linear relationship exists between Darcy permeability and the reciprocal of mean pressure in the system, that is, between gas-flux reduction and mean pressure increase.

$$k(p_{avg}) = k_D \left(1 + \frac{b}{p_{avg}} \right), \tag{2.13}$$

where k (p_{avg}) is the gas permeability at mean pressure (p_{avg}), k_D the Darcy permeability or liquid permeability, and b the Klinkenberg parameters. The empirical parameters b and k_D are the slope and intercept of the fitted line through the k (p_{avg}) versus $\frac{1}{p_{avg}}$ data. The Klinkenberg effect has been used to model the gas flow in conventional gas reservoirs (with pores in the range of 10–100 μm) and recently for tight gas systems (with pores of 1–10 μm in size).

The flow of gas in micro- or nanochannels can be described by use of molecular models, commonly known as molecular dynamics, which consider the molecular nature of a gas (Gad-el-Hak 1999) or Lattice-Boltzmann method (Shabro et al. 2012). Although these molecular models are valid for any range of K_n, the requirement of large computational time and power constitutes a major limitation of these approaches, currently rendering them unfeasible for shale analysis. A common way to model the flow of gas through micro- or nanochannels is to modify the no-slip boundary condition in continuum models by accounting for a slip boundary condition. This approach has been used in multiple proposed models for shale gas transport (Javadpour 2009; Civan 2010; Azom and Javadpour 2012; Darabi et al. 2012).

Javadpour (2009) proposed a model that includes the Knudsen diffusion and the slip flow which are major mechanisms contributing to the gas flow in a single, straight, cylindrical nanotube. Javadpour also asserted that these two processes exist at any K_n, but their individual contributions to total flux varies. Javadpour (2009) proposed a model for gas flow in a nanopore duct by accounting for Knudsen diffusion and slip velocity using the Maxwell theory.

$$J = \left[\frac{2rM}{3 \times 10^3 RT} \left(\frac{8RT}{\pi M} \right)^{0.5} + F \frac{r^2 \rho_{avg}}{8\mu} \right] \frac{p_2 - p_1}{L} \qquad (2.14)$$

where J is the mass flux or molar flux, r the pore radius, M the molar mass, F the slip coefficient, ρ_{avg} the average density, and L the length of the media. p_1 and p_2 are the upstream and downstream pressures respectively. The first and second terms in the right-hand-side bracket in Eq. 2.14 refer to Knudsen diffusion and slip flow, respectively. The term F is the slip coefficient and is defined as:

$$F = 1 + \left(\frac{8\pi RT}{M} \right)^{0.5} \frac{\mu}{r p_{avg}} \left(\frac{2}{\alpha} - 1 \right) \qquad (2.15)$$

where the α is the tangential momentum accommodation coefficient or the fraction of gas molecules reflected diffusely from the pore wall relative to specular reflection. The value of α varies theoretically in a range from 0 (representing specular accommodation) to 1 (representing diffuse accommodation), depending on wall-surface smoothness, gas type, temperature, and pressure (Agrawal and Prabhu 2008; Arkilic et al. 2001). Experimental measurements are needed to determine α for specific shale systems.

Javadpour (2009) showed that this model matches data by Roy et al. (2003), from flow through a membrane with pore sizes of 200 nm, at an average error of 4.5 %. By comparing Eq. 2.14 to Darcy's law for a single nanotube (Hagen-Poiseuille equation), apparent permeability, k_{app}, for a porous medium containing of straight cylindrical nanotubes can be defined as:

$$k_{app} = \frac{2r\mu}{3 \times 10^3 p_{avg}} \left(\frac{8RT}{\pi M} \right)^{0.5} + \frac{r^2}{8} \left\{ 1 + \left(\frac{8\pi RT}{M} \right)^{0.5} \left(\frac{2}{\alpha} - 1 \right) \frac{\mu}{r p_{avg}} \right\} \qquad (2.16)$$

where k_{app} apparent permeability. Equation 2.16 provides an apparent Darcy permeability relationship written in the Klinkenberg form as

$$k_{app} = k_D \left(1 + \frac{b}{p_{avg}} \right) \tag{2.17}$$

$$b = \frac{16\mu}{3 \times 10^3 r} \left(\frac{8RT}{\pi M} \right)^{0.5} + \left(\frac{8\pi RT}{M} \right)^{0.5} \left(\frac{2}{\alpha} - 1 \right) \frac{\mu}{r}, \tag{2.18}$$

Azom and Javadpour (2012) showed how Eq. 2.16 can be corrected for a real gas flowing in a porous medium. The final equation still has the form of Eq. 2.17, but with b given below

$$b = \frac{16\mu c_g p_{avg}}{3 \times 10^3 r} \left(\frac{8ZRT}{\pi M} \right)^{0.5} + \left(\frac{8\pi RT}{M} \right)^{0.5} \left(\frac{2}{\alpha} - 1 \right) \frac{\mu}{r}, \tag{2.19}$$

where c_g is gas compressibility and Z is compressibility factor. Notice that as the real gas becomes ideal, Eq. 2.19 becomes Eq. 2.18, because the gas compressibility $c_g = \frac{1}{p_{avg}}$ and the compressibility factor $z = 1$ for an ideal gas.

Darabi et al. (2012) later applied several modifications to adapt the model developed by Javadpour (2009) from being applicable to a single, straight, cylindrical nanotube to being applicable to ultra-tight, natural porous media characterized by a network of inter-connected tortuous micropores and nanopores. This model accounts for Knudsen diffusion and surface roughness, in addition to slip flow, by use of the Maxwell theory.

$$k_{app} = \frac{\mu M \phi}{RT \tau \rho_{avg}} (\delta_r)^{D_f - 2} D_k + k_D \left(1 + \frac{b}{p_{avg}} \right). \tag{2.20}$$

In Eq. 2.20, τ is the tortuosity and δ_r the ratio of normalized molecular radius size, r_m, with respect to local average pore radius, r_{avg}, yielding $\delta_r = \frac{r_m}{r_{avg}}$. In above equation, Knudsen diffusion coefficient, D_k, is defined as

$$D_k = \frac{2r_{avg}}{3} \left(\frac{8RT}{\pi M} \right)^{0.5}, \tag{2.21}$$

where r_{avg} is approximated by $r_{avg} = (8k_D)^{0.5}$. The average pore radius can also be determined by laboratory experiments employing such as processes as mercury injection and nitrogen adsorption tests and pore imaging using SEM and AFM.

Darabi et al. (2012) also included the fractal dimension of the pore surface, D_f, to consider the effect of pore-surface roughness on the Knudsen diffusion coefficient (Coppens 1999; Coppens and Dammers 2006). Surface roughness is one example of local heterogeneity. Increasing surface roughness leads to an increase in residence time of molecules in porous media and a decrease in Knudsen diffusivity. D_f is a quantitative measure of surface roughness that varies between 2 and 3, representing a smooth surface and a space-filling surface, respectively (Coppens and Dammers 2006).

Civan (2010) permeability model is based on the Beskok and Karniadakis (1999) approach. The model represented by simplified second-order slip approach assumes that permeability is a function of the intrinsic permeability, the Knudsen number, K_n, the rarefaction coefficient α_r, and the slip coefficient b,

$$k = k_D(1 + \alpha_r K_n)\left(1 + \frac{4K_n}{1 - bK_n}\right). \tag{2.22}$$

The dimensionless rarefaction coefficient α_r is given by,

$$\alpha_r = \alpha_0\left(\frac{K_n^B}{A + K_n^B}\right). \tag{2.23}$$

where A and B are empirical fitting constants. The lower limit of $\alpha_r(\alpha_r = 0)$ corresponds to the slip flow regime and the upper limit α_0 corresponds to the asymptotic limit of α_r when $K_n \rightarrow \infty$, which corresponds to the free molecular flow. Constants A and B serve as the fitting parameters that may be appropriately adjusted based on the dominant flow regime in the shale porous media. Civan (2010) reports the adjusted parameter values, $A = 0.178$, $B = 0.4348$, and $\alpha_0 = 0.1358$ for modeling gas flow in a tight sand example. Civan (2010) assumes $b = -1$ based on the Beskok and Karniadakis (1999) estimate and subsequently estimates the Knudsen number as (Jones and Owens 1980),

$$K_n = 12.639k_D^{-1/3}. \tag{2.24}$$

With these assumptions, the only unknown parameter remaining in the Civan (2010) model is k_D, which can be determined from a permeability measurement experiment (e.g., the pulse-decay experiment).

For small Knudsen numbers, that is, $K_n \ll 1$, Civan (2010) estimates the dynamic slippage coefficient b_k as a function of gas viscosity, based on the Florence et al. (2007)'s study,

$$b_k = \frac{2790\mu}{\sqrt{M}}\left(\frac{k_D}{\phi}\right)^{-0.5}. \tag{2.25}$$

The most important limitation to these discussed models is the estimation of empirical parameters which requires performing experiments or computationally expensive molecular-dynamic simulations (Agrawal and Prabhu 2008). Singh et al. (2014) proposed a new non-empirical, analytical model for permeability, termed non-empirical apparent permeability (NAP). NAP is developed for flow of gas in ultra-tight porous media consisting of tortuous micro-/nano-pores and is valid for Knudsen numbers less than unity and stands up under the complete operating conditions of shale reservoirs.

Singh et al. (2014) derived apparent permeability on the basis of fundamental flow equations for shale-gas systems. From the total mass flow which is a super-position of advection and molecular spatial diffusion (Veltzke and Thöming 2012), Darcy's law can be converted to expressions for apparent permeability of slits or tubes:

$$(k_{\text{app}})_{\text{slit}} = \frac{\phi\mu h}{3\tau}\left(\frac{h_{\text{slit}}Z}{4\mu} \frac{8}{\pi p_{\text{avg}}M}\sqrt{\frac{2MRT}{\pi}}\right) \tag{2.26}$$

$$(k_{\text{app}})_{\text{tube}} = \frac{2\phi\mu d}{\pi\tau}\left(\frac{\pi d_{\text{tube}}Z}{64\mu} \frac{1}{3p_{\text{avg}}M}\sqrt{2\pi MRT}\right) \tag{2.27}$$

where h_{slit} is the height of the rectangular slit and d_{tube} the diameter of the tube. The two pore geometries considered in the NAP model are cylindrical tube and rect-angular channel (slit). When porous media are composed of other shapes, the permeability of the media will be somewhere between what it would be if it were composed of tubes and what it would be if it were composed of slits. Therefore, the two shapes considered in the NAP model may reliably capture the average effect of different pore shapes in porous media because capturing the exact shape of each pore might be impractical and daunting. The permeability of each shape type contributes to the effective permeability of the reservoir, where the effective per-meability is the statistical sum of the individual permeability from each shape type (Fenton 1960) as given below:

$$\ln(k_{\text{app}})_{\text{eff}} = \frac{x}{100}\ln(k_{\text{app}})_{\text{slit}} + \frac{100-x}{100}\ln(k_{\text{app}})_{\text{tube}} \tag{2.28}$$

$$(k_{\text{app}})_{\text{eff}} = \left[\left(k_{\text{app}}^{\frac{x}{100}}\right)_{\text{slit}}\left(k_{\text{app}}^{\frac{100-x}{100}}\right)_{\text{tube}}\right] \tag{2.29}$$

where k_{eff} is effective permeability after including the effect of sorption. The novelty of this work is the development of flow equations without empirical parameters. Although there are some empirical values of simple gases and solid materials in literature, finding them for the shale system is not straightforward. Hence, a method that does not need the empirical value is attractive.

Figure 2.7 compares the predictions of cumulative gas production for the APF (Darabi et al. 2012), NAP (Singh et al. 2014), Klinkenberg (1941), Civan (2010), Darcy-type-flow, and Knudsen-diffusion models (Javadpour 2009). The NAP pre-dictions lie between APF and Klinkenberg, whereas Civan and Klinkenberg pre-dictions are close to each other and each of them is higher than the predictions by Darcy. At the given typical shale-gas-reservoir conditions, contribution of Darcy flow, slip flow, and Knudsen diffusion control total gas production. The APF model

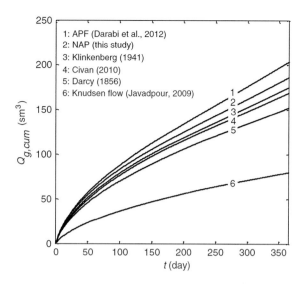

Fig. 2.7 Comparison of different gas models to predict cumulative gas production (Singh et al. 2014)

includes all these three processes, whereas the NAP model ignores slip flow. A comparison between APF (Darabi et al. 2012) and NAP model in Fig. 2.7 suggests that the Klinkenberg effect is not dominant at high-Knudsen-number flow (applicable to shale gas) and that a combination of Darcy-type flow corrected for Knudsen diffusion can be used alternatively.

As mentioned earlier in this chapter, gas storage in gas shale exists in three major forms: stored as compressed gas in the pore network, sorbed on the surface of organic material and possibly on clay minerals, and dissolved in liquid hydro-carbon and brine (interstitial and clay-bound), and kerogen (Javadpour et al. 2007). Many research studies have addressed the first two storage processes (Chareonsuppanimit et al. 2012; Civan et al. 2012; Darabi et al. 2012; Javadpour 2009; Zhang et al. 2012), but only limited research has been conducted on the contribution of gas dissolved in organic material in the total gas production from shale reservoirs (Etminan et al. 2014; Moghanloo et al. 2013).

Figure 2.8 shows the gas-molecule in a part of pore system including kerogen. The compressed gas exists in the micro- and nano-scale pores. Some of the gas molecules are adsorbed on the surface of kerogen and, eventually, some of the gas molecules are dissolved into the kerogen body and become a part of the kerogen in the form of a single phase. The controlling mass transport process of the dissolved gas is molecular diffusion. Depending on the geochemistry of the organic materials (thermal maturity, organic source, etc.), different gas solubility could be expected. The contribution of dissolved gas to gas-in-place and ultimate recovery of a shale reservoir could be significant; hence, evaluation of the gas-diffusion process into kerogen becomes important. In addition to the total contribution of each process, the onset time of each process during production is critical. Once production starts from a reservoir, the compressed gas in interstitial pore spaces expands first; then, adsorbed gas on the surfaces of the pores in kerogen desorbs to the pore network.

Fig. 2.8 Schematic view of gas-molecule locations in a small part of pore system including kerogen (Javadpour 2009)

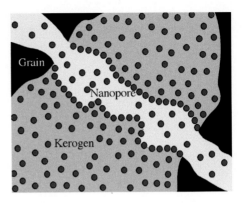

At this stage, the concentration of gas molecules on the pore inner surface decreases and creates a concentration gradient in the bulk of the kerogen, thereby triggering gas diffusion (Etminan et al. 2014; Javadpour et al. 2007).

2.5   Non-Darcy Flow

In 1856, Darcy developed his now famous flow correlation by flowing water, available at the local hospital, through sand pack configurations. Darcy's law, shown in Eq. 2.30, describes the linear proportionality involving a constant, k, as related to the potential gradient $\frac{dp}{dx}$, the fluid viscosity of μ, and the superficial velocity of v.

$$-\frac{dp}{dx} = \frac{\mu v}{k} \qquad (2.30)$$

where v is the superficial velocity. Forty-five years later, Forchheimer (1901) observed deviation from the linearity of Darcy's equation at increased flow rates. When the gas velocity increases, for example near the drain inside hydraulic fractures, significant inertial (non-Darcy) effects can occur. This induces an additional pressure drop in the hydraulic fractures in order to maintain the production rate. Forchheimer proposed a second proportionality constant, in addition to k, that would account for this non-linearity. He called this second proportionality constant, β, and it resulted in the familiar Forchheimer equation shown in Eq. 2.31.

$$-\frac{dp}{dx} = \frac{\mu v}{k} + \beta \rho v^2 \qquad (2.31)$$

where β is the non-Darcy flow coefficient. The earliest references to non-Darcy flow effects in petroleum literature occur in the early 1960s (Carter 1962; Swift and

Kiel 1962; Tek et al. 1962). The effects of non-Darcy flow specifically in hydraulic fracturing operations were first addressed by Cooke (1973) as given below:

$$\beta = bk^{-a} \tag{2.32}$$

where a and b are constant determined by experiments based on proppant type. Equation 2.32 is simple and applicable to different types of proppants.

Geertsma (1974) developed a dimensionally consistent correlation between the non-Darcy flow coefficient, permeability and porosity. Analyzing the data obtained for unconsolidated sandstones, consolidated sandstones, limestones, and dolomites from his and other experiments (Green and Duwez 1951; Cornell and Katz 1953) and performed dimensional analysis, he reached an empirical correlation,

$$\beta = \frac{0.005}{\phi^{5.5} k^{0.5}}. \tag{2.33}$$

In addition to the one phase correlation Eq. 2.33, Geertsma (1974) proposed a correlation for β in a two-phase system. He argued that, in the two-phase system, the permeability in Eq. 2.33 would be replaced by the gas effective permeability at a certain water saturation, while the porosity would be replaced by the void fraction occupied by the gas. Therefore, in the two-phase system, where the fluid was immobile, the β correlation became

$$\beta = \frac{0.005}{\phi^{5.5} k^{0.5}} \left[\frac{1}{(1 - S_{wr})^{5.5} k_r^{0.5}} \right]. \tag{2.34}$$

where S_{wr} is the residual water saturation and k_r is the relative permeability. Equation 2.34 shows that the presence of the liquid phase increases the non-Darcy coefficient.

Evans and Civan (1994) presented a general correlation for the non-Darcy flow coefficient using a large variety of data from consolidated and unconsolidated media including the effects of multiphase fluids and overburden stress. They collected a total of 183 data points and also employed data from Geertsma (1974) in consolidated media, and from Evans and Evans (1988) for the effects of immobile liquid saturation and closure stress on the β-coefficient in propped fractures. The regression line yielded the following general correlation:

$$\beta = \frac{1.485 \times 10^9}{\phi k^{1.021}} \tag{2.35}$$

with correlation coefficient $R = 0.974$. Since this correlation is obtained from a large variety of porous media under different conditions, it is expected to provide a reasonable estimation for the β-coefficient. Equation 2.35 is implemented in the numerical model and used for accounting for non-Darcy flow in hydraulic fractures.

In addition to this correlation, there are several theoretical and empirical correlations of the non-Darcy coefficient in literatures and these are reviewed by Evans and Civan (1994) and Dacun and Thomas (2001).

2.6   Stress-Dependent Compaction

In shale formations, the conductivity of fracture network is sensitive to the change in stress and strain during production because natural fractures are weakly propped compared with hydraulic fractures. Figure 2.9 shows that experimental results measuring permeability and porosity with respect to effective confining pressure (Dong et al. 2010). Therefore, geomechanical effects during production must be included to simulate the stress-dependent effects of shale gas reservoir.

Previous researches successfully proved that iterative coupling between geomechanics and reservoir flow allows easy control of convergence as well as easy maintenance of the reservoir and geomechanics simulators (Tran et al. 2005, 2010). However, it is concluded that linear elastic model cannot solely describe shale gas reservoirs (Li and Ghassemi 2012; Hosseini 2013). In order to consider the change of conductivity, pressure-dependent properties were presented in several researches (Pedrosa 1986; Raghavan and Chin 2004; Cho et al. 2013). Therefore, the deformation of shale reservoir should be modeled by stress-dependent correlations coupled with linear-elastic model. To consider decreasing production caused by porosity and permeability reduction in the shale gas reservoir model, stress-dependent porosity and permeability correlations are applied with a linear elastic constitutive model. Dong et al. (2010) used exponential and power law correlations to match the experimental data as follows:

$$\phi = \phi_i e^{-a(\sigma' - \sigma_i')} \tag{2.36}$$

$$k = k_i e^{-b(\sigma' - \sigma_i')} \tag{2.37}$$

$$\phi = \phi_i \left(\frac{\sigma'}{\sigma_i'}\right)^{-c} \tag{2.38}$$

$$k = k_i \left(\frac{\sigma'}{\sigma_i'}\right)^{-d} \tag{2.39}$$

where σ' is the effective stress and a, b, c, and d are experimental coefficients. The subscript i indicates the initial state. Figure 2.10 shows that results of curve fitting on measured porosity and permeability of the shale cores with exponential and power law correlations (Dong et al. 2010).

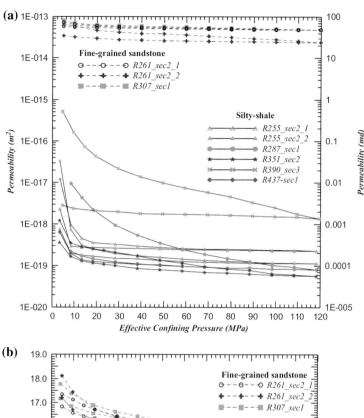

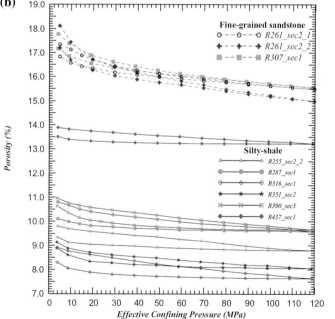

Fig. 2.9 Stress-dependent **a** permeability and **b** porosity of the sandstone (*red dashed lines*) and silty-shale (*solid black lines*) (Dong et al. 2010)

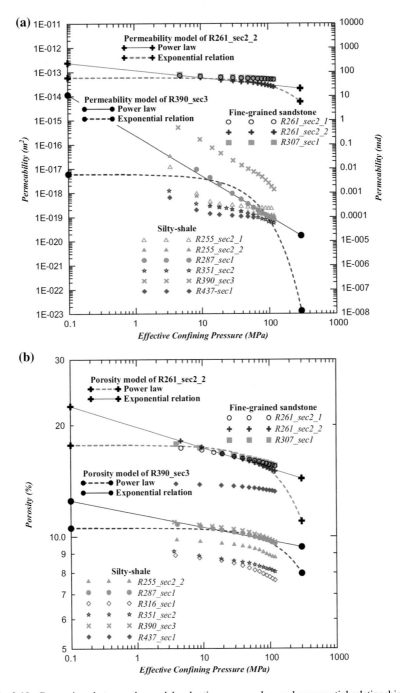

Fig. 2.10 Comparison between the models adopting a power law and exponential relationship for **a** permeability and **b** porosity (Dong et al. 2010)

References

Agrawal A, Prabhu SV (2008) Survey on measurement of tangential momentum accommodation coefficient. J Vac Sci Technol A 26(4):634–645. doi:10.1116/1.2943641

Arkilic EB et al (2001) Mass flow and tangential momentum accommodation in silicon micromachined channels. J Fluid Mech 437:29–43

Azom P, Javadpour F (2012) Dual-continuum modeling of shale and tight gas reservoirs. Paper presented at the SPE annual technical conference and exhibition, San Antonio, Texas, 8–10 Oct 2012. doi:10.2118/159584-MS

Barenblatt GI et al (1960) Basic concept in the theory of seepage of homogeneous liquids in fissured rocks. J Appl Math Mech 24(5):1286–1303

Beskok A, Karniadakis GE (1999) A model for flows in channels, pipes, and ducts at micro and nano scales. Microsc Therm Eng 3(1):43–77. doi:10.1080/108939599199864

Brown GP et al (1946) The flow of gases in pipes at low pressures. J Apple Phys 17:802–813

Brunauer S et al (1938) Adsorption of gases in multimolecular layers. J Am Chem Soc 60(2):309–319

Carter RD (1962) Solutions of unsteady-state radial gas flow. J Pet Tech 14(05):549–554. doi:10.2118/108-PA

Chareonsuppanimit P et al (2012) High-pressure adsorption of gases on shales: measurements and modeling. Int J Coal Geol 95:34–46. doi:10.1016/j.coal.2012.02.005

Cho Y et al (2013) Pressure-dependent natural-fracture permeability in shale and its effect on shale-gas well production. SPE Res Eval Eng 16(2):216–228. doi:10.2118/159801-PA

Civan F (2010) Effective correlation of apparent gas permeability in tight porous media. Transp Porous Med 82(2):375–384. doi:10.1007/s11242-009-9432-z

Cooke CE (1973) Conductivity of fracture proppants in multiple layers. J Pet Tech 25(09):1101–1107. doi:10.2118/4117-PA

Coppens M-O (1999) The effect of fractal surface roughness on diffusion and reaction in porous catalysts from fundamentals to practical applications. Catal Today 53(2):225–543. doi:10.1016/S0920-5861(99)00118-2

Coppens M-O, Dammers AJ (2006) Effects of heterogeneity on diffusion in nanopores from inorganic materials to protein crystals and ion channels. Fluid Phase Equilibr 241 (1–2):308–316. doi:10.1016/j.fluid.2005.12.039

Cornell D, Katz DL (1953) Flow of gases through consolidated porous media. Ind Eng Chem 45 (10):2145–2152. doi:10.1021/ie50526a021

Dacun L, Thomas WE (2001) Literature review on correlations of the non-darcy coefficient. Paper presented at the SPE Permian basin oil and gas recovery conference, Midland, Texas, 15–17 May 2001. doi:org/10.2118/70015-MS

Darabi H et al (2012) Gas flow in ultra-tight shale strata. J Fluid Mech 710:641–658. doi:10.1017/jfm.2012.424

Dong JJ et al (2010) Stress-dependence of the permeability and porosity of sandstone and shale from TCDP Hole-A. Int J Rock Much Min Sci 47(7):1141–1157. doi:10.1016/j.ijrmms.2010.06.019

Etminan SR et al (2014) Measurement of gas storage processes in shale and of the molecular diffusion coefficient in kerogen. Int J Coal Geol 123:10–19. doi:10.1016/j.coal.2013.10.007

Evans EV, Evans RD (1988) The influence of an immobile or mobile saturation on non-Darcy compressible flow of real gases in propped fractures. J Pet Tech 1345–1351. doi:10.2118/15066-PA

Evans RD, Civan F (1994) Characterization of non-darcy multiphase flow in petroleum bearing formation. U.S. Department of Energy, Washington, D.C

Fenton L (1960) The sum of log-normal probability distributions in scatter transmission systems. IEEE T Commun 8(1):57–67. doi:10.1109/TCOM.1960.1097606

Florence FA et al (2007) Improved permeability prediction relations for low-permeability sands. Paper presented at SPE Rocky mountain oil and gas technology symposium, Denver, Colorado, 16–18 April 2007

Forchheimer P (1901) Wasserbewegung durch boden. Zeits V Deutsch Ing 45:1781–1901

Freeman CM et al (2012) Measurement, modeling, and diagnostics of flowing gas composition changes in shale gas wells. Paper presented at the SPE Latin American and Caribbean petroleum engineering conference, Mexico City, Mexico, 16–18 April 2012

Gad-el-Hak M (1999) The fluid mechanics of microdevices—the freeman scholar lecture. J Fluids Eng 121(1):5. doi:10.1115/1.2822013

Gao C et al (1994) Modeling multilayer gas reservoirs Including sorption effects. Paper presented at the SPE eastern regional conference and exhibition, Charleston, West Virginia, 8–10 Nov 1994

Geertsma J (1974) Estimating the coefficient of inertial resistance in fluid flow through porous media. Soc Pet Eng J 14(05):445–450. doi:10.2118/4706-PA

Green L, Duwez PJ (1951) Fluid flow through porous metals. J Appl Mech 18(1):39

Hosseini SM (2013) On the linear elastic fracture mechanics application in Barnett shale hydraulic fracturing. Paper presented at the 47th U.S. rock mechanics/geomechanics symposium, San Francisco, California, 23–26 June 2013

Javadpour F (2009) Nanopores and apparent permeability of gas flow in mudrocks (shales and siltstone). J Can Pet Tech 48(8):16–21. doi:10.2118/09-08-16-DA

Javadpour F et al (2007) Nanoscale gas flow in shale gas sediments. J Can Pet Tech 46(10):55–61. doi:10.2118/07-10-06

Jones FO, Owens WW (1980) A laboratory study of low-permeability gas sands. J Pet Technol 32 (9):1631–1640

Klinkenberg LJ (1941) The permeability of porous media to liquids and gases. In: Drilling and production practice, New York, New York, January 1941

Kuila U, Prasad M (2013) Specific surface area and pore-size distribution in clays and shales. Geophys Prosp 61(2):341–362

Langmuir I (1918) The adsorption of gases on plane surfaces of glass, mica and platinum. J Am Chem Soc 40:1403–1461

Li Y, Ghassemi A (2012) Creep behavior of Barnett, Haynesville, and Marcellus shale. Paper presented at the 46th U.S. rock mechanics/geomechanics symposium, Chicago, Illinois, 24–27 June 2012

Moghanloo RG et al (2013) Contribution of methane molecular diffusion in kerogen to gas-in-place and production. Paper presented at the SPE western regional and AAPG pacific section meeting 2013 Joint technical conference, Monterey, California, 19–25 April. doi:10. 2118/165376-MS

Pedrosa OA (1986) Pressure transient response in stress-sensitive formations. Paper presented at the SPE California regional meeting, Oakland, California, 2–4 April 1986. doi:10.2118/15115-MS

Raghavan R, Chin LY (2004) Productivity changes in reservoirs with stress-dependent permeability. SPE Res Eval Eng 7(4):308–315. doi:10.2118/88870-PA

Rathakrishnan E (2004) Gas dynamics. Prentice-hall of India Pvt Ltd, New Delhi, India

Rezaee R (eds) (2015) Fundamental of gas shale reservoirs. Wiley, New Jersey

Roy S et al (2003) Modeling gas flow through microchannels and nanopores. J Appl Phys 93:4870–4879. doi:10.1063/1.1559936

Shabro V et al (2012) Finite-difference approximation for fluid-flow simulation and calculation of permeability in porous media. Transport Porous Med 94(3):775–793. doi:10.1007/s11242-012-0024-y

Silin D, Kneafsey T (2012) Shale gas: nanometer-scale observations and well modeling. J Can Pet Tech 51(6):464–475

Sing KSW et al (1985) Reporting physisorption data for gas/solid systems with special reference to the determination of surface area and porosity. Pure Appl Chem 57(4):603–619

Singh H et al (2014) Nonempirical apparent permeability of shale. SPE Res Eval Eng 17(3):414–424. doi:10.2118/170243-PA

Stewart G (2011) Well test design and analysis. Pennwell, Tulsa, Oklahoma

Swift GW, Kiel OG (1962) The prediction of gas-well performance including the effect of non-darcy flow. J Pet Tech 14(07):791–798. doi:10.2118/143-PA

Tek MR et al (1962) The effect of turbulence on flow of natural gas through porous reservoirs. J Pet Tech 14(07):799–806. doi:10.2118/147-PA

Tran D et al (2005) An overview of iterative coupling between geomechanical deformation and reservoir flow. Paper presented at the SPE international thermal operations and heavy oil symposium, Calgary, Alberta, Canada, 1–3 Nov 2005. doi:10.2118/97879-MS

Tran D et al (2010) Improved gridding technique for coupling geomechanics to reservoir flow. Soc Pet Eng J 15(1):64–75. doi:10.2118/115514-PA

Veltzke T, Thöming J (2012) An analytically predictive model for moderately rarefied gas flow. J Fluid Mech 698:406–422. doi:10.1017/jfm.2012.98

Warren JE, Root PJ (1963) The behavior of naturally fractured reservoirs. Soc Petrol Eng J 3(3):245–255

Yu W et al (2014) Evaluation of gas adsorption in Marcellus shale. Paper presented at the SPE annual technical conference and exhibition, Amsterdam, The Netherlands, 27–29 Oct 2014

Zhang T et al (2012) Effect of organic-matter type and thermal maturity on methane adsorption in shale gas systems. Org Geochem 47:120–131. doi:10.1016/j.orggeochem.2012.03.012

Chapter 3
Numerical Modeling

3.1 Introduction

Realistic modeling of shale gas reservoir is an important issue in these days. For the accurate modeling of shale reservoir, distinguishing features of shale should be considered. Natural fracture system can be simplified to dual porosity and dual permeability models. These models present the system as an orthogonal set of intersecting fractures and cubic matrix blocks. Adsorption of hydrocarbon gas in the matrix surface is also considered with Langmuir isotherm. Non-Darcy flow in the fractures caused by turbulent flow is computed with Forchheimer equation. In other to consider deformation of shale rock, stress and strain are calculated by geomechanical model and stress-dependent correlations are used to compute permeability and porosity of shale gas reservoir. Based on these mechanisms, synthetic numerical model is presented preferentially. Finally, verification of shale gas reservoir model with Barnett field data is provided. For applying effect of stress-dependent compaction, experimental coefficients are estimated by correlation between exponential and power law coefficients. Results of history matching are shown with and without considering stress-dependent compaction. Furthermore, results of history matching with concept of SRV are presented.

3.2 Modeling of Shale Gas Reservoir

A number of studies have presented the numerical model of shale reservoirs (Cipolla et al. 2010; Rubin 2010; Yu et al. 2013; Lee et al. 2014; Kim et al. 2014, 2015; Kim and Lee 2015). For realistic modeling of shale gas reservoir, natural

© The Author(s) 2016
K.S. Lee and T.H. Kim, *Integrative Understanding of Shale Gas Reservoirs*,
SpringerBriefs in Applied Sciences and Technology,
DOI 10.1007/978-3-319-29296-0_3

fracture system, multi-fractured horizontal well, adsorption of methane, non-Darcy flow, stress-dependent compaction, and stimulated reservoir volume (SRV) should be considered.

Dual porosity and dual permeability model can be used for simulate natural fracture system. These reservoir models show difference in methods of matrix and fractures description in a shale gas reservoir. In dual porosity model which was presented by Warren and Root (1963), fractures are the only pathway connected to the wellbore. The matrix of dual porosity system is not connected to the wellbore directly and the fluid of the matrix is transported to the well through the fractures. Dual permeability system is similar to the dual porosity system except that matrix blocks of dual permeability system have one more channel for fluids flow than those of dual porosity system. Dual permeability system assumes that both matrix and fractures are connected to the wellbore directly. The fluid could flow from the fracture and matrix to the wellbore as well as travel between the matrix and fractures at the same time.

The following describes the governing equations for the dual porosity and dual permeability approach to modeling naturally fractured reservoirs (CMG 2015). The governing equations of dual porosity model are an extension of the equations for single porosity systems. The representation of the matrix follows Kazemi et al. (1978) where fractures are assumed orthogonal in the three directions and acts as boundaries to matrix elements. Dual porosity formulations in matrix (Eqs. 3.1 and 3.2) and fracture (Eqs. 3.3 and 3.4) blocks are given below

$$\psi_{im} = -\tau_{iomf} - \tau_{igmf} - \frac{V}{\Delta t}\left(N_i^{n+1} - N_i^n\right)_m = 0, \quad i = 1, \ldots, n_c \tag{3.1}$$

$$\psi_{n_c+1,m} = -\tau_{wmf} - \frac{V}{\Delta t}\left(N_{n_c+1}^{n+1} - N_{n_c+1}^n\right)_m = 0 \tag{3.2}$$

$$\psi_{if} = \Delta T_{of}^s y_{iof}^s \left(\Delta p^{n+1} - \gamma_o^s \Delta D\right)_f + \Delta T_{gf}^s y_{igf}^s \left(\Delta p^{n+1} + \Delta p_{cog}^s - \gamma_g^s \Delta D\right)_f$$
$$+ q_i^{n+1} + \tau_{iomf} + \tau_{igmf} - \frac{V}{\Delta t}\left(N_i^{n+1} - N_i^n\right) = 0, \quad i = 1, \ldots, n_c \tag{3.3}$$

$$\psi_{n_c+1,f} = \Delta T_{wf}^s \left(\Delta p^{n+1} - \Delta p_{cwo}^s - \gamma_w^s \Delta D\right)_f + q_w^{n+1} + \tau_{wmf}$$
$$- \frac{V}{\Delta t}\left(N_{n_c+1}^{n+1} - N_{n_c+1}^n\right)_f$$
$$= 0 \tag{3.4}$$

where ψ is the material balance equation, τ_{iomf} the matrix-fracture transfer in the oil phases for component i, τ_{igmf} the matrix-fracture transfer in the gas phases for component i, τ_{wmf} the matrix-fracture transfer for water, V the grid block volume, Δt the time step, N_i the moles of component i per unit of grid block volume, N_{n_c+1} the moles of water per unit of grid block volume, T_j the transmissibility of phase j, y_{ij} the mole fraction of component i in phase j, γ_j the gradient of phase j, D the depth, p_{cog} the oil-gas capillary pressure, and p_{cwo} the water-oil capillary pressure. The subscript i with $i = 1, \ldots, n_c$ corresponds to the hydrocarbon component and the subscript $n_c + 1$ denotes the water component. Subscripts j indicates phase of oil, gas, and water, presented by o, g, and w. The superscripts n and $n + 1$ denote respectively the old and current time level and the superscript s refers to n for explicit blocks and to $n + 1$ for implicit blocks. The subscripts f and m correspond to the fracture and matrix respectively.

Dual permeability formulations are similar to the dual porosity formulation, except that matrix blocks are connected to one another and thus provide alternate channels for fluid flow. The fracture equations are the same as those in dual porosity formulation. The matrix flow equations contain additional terms as follows:

$$\psi_{im} = \Delta T^s_{om} y^s_{iom} \left(\Delta p^{n+1} - \gamma^s_o \Delta D \right)_m + \Delta T^s_{gm} y^s_{igm} \left(\Delta p^{n+1} + \Delta p^s_{cog} - \gamma^s_g \Delta D \right)_m$$
$$- \tau_{iomf} - \tau_{igmf} - \frac{V}{\Delta t} \left(N^{n+1}_i - N^n_i \right)_m = 0 \quad i = 1, \ldots, n_c \tag{3.5}$$

$$\psi_{n_c+1,m} = \Delta T^s_{wm} \left(\Delta p^{n+1} - \Delta p^s_{cwo} - \gamma^s_w \Delta D \right)_m - \tau_{wmf} - \frac{V}{\Delta t} \left(N^{n+1}_{n_c+1} - N^n_{n_c+1} \right)_m = 0 \tag{3.6}$$

There are several methods for calculating matrix-fracture transfer and one of them considering pseudo capillary pressure and partially immersed matrix are presented as follows

$$\tau_{omf} = \sigma V \frac{k_{ro} \rho_o}{\mu_o} \left(p_{om} - p_{of} \right) \tag{3.7}$$

$$\tau_{gmf} = \sigma V \frac{k_{rg} \rho_g}{\mu_g} \left\{ \left(p_{om} - p_{of} \right) + \left[S_{gm} + \frac{\sigma_z}{\sigma} \left(\frac{1}{2} - S_{gm} \right) \right] \left(\tilde{p}_{cog,m} - \tilde{p}_{cog,f} \right) \right\} \tag{3.8}$$

$$\tau_{wmf} = \sigma V \frac{k_{rw} \rho_w}{\mu_w} \left\{ \begin{array}{l} \left(p_{om} - p_{of} \right) - \left(p_{cwo,m} - p_{cwo,f} \right) \\ - \left(\frac{1}{2} \frac{\sigma_z}{\sigma} \right) \left[\left(\tilde{p}_{cwo,m} - \tilde{p}_{cwo,f} \right) - \left(p_{cwo,m} - p_{cwo,f} \right) \right] \end{array} \right\} \tag{3.9}$$

where σ is the transfer coefficient.

As shown in Eq. 3.10 below, Langmuir isotherm is considered to describe the adsorption capacity of rock as a function of pressure changes under isothermal condition.

$$V = \frac{V_L p}{p + p_L}, \tag{3.10}$$

where V is the gas volume of adsorption at pressure p, the Langmuir volume V_L indicates the maximum gas volume can be adsorbed and the Langmuir pressure p_L is the pressure at which half of Langmuir volume gas is stored.

For realistic modeling flow in hydraulic fractures, non-Darcy effect should be considered. Forchheimer equation (1901) developed for inertial effect are shown as next

$$-\frac{dp}{dx} = \frac{\mu v}{k} + \beta \rho v^2 \qquad (3.11)$$

where β the non-Darcy flow coefficient or Forchheimer β coefficient. To compute non-Darcy flow coefficient, Evans and Civan's (1994) empirical correlation can be used given below.

$$\beta = \frac{1.485 \times 10^9}{\phi k^{1.021}} \qquad (3.12)$$

Gas slippage is a phenomenon associated with non-laminar gas flow effects in porous media. At low pressure, the velocity of the individual gas molecules tends to accelerate or slip along the pore wall of porous medium. As a consequence, permeability can be overestimated without considering this gas slippage effect. This phenomenon is so called Klinkenberg effect (1941) and especially significant in low permeability or shale gas reservoirs characterized by small pore throat. Darcy's law requires a correction for the mean flowing pressure. Effective gas permeability at a specific pressure is given by

$$k_g = k_D \left(1 + \frac{b}{p_{\text{avg}}} \right) \qquad (3.13)$$

In shale gas reservoir, hydraulic fracturing not only creates new fractures but also rejuvenates existing natural fractures, which opens networks of interconnected fractures around the wellbore. Kim and Lee (2015) distinguished rejuvenated fractures and natural fractures to construct more accurate shale gas reservoir. Rejuvenated fractures which compose SRV with hydraulic fractures are formed near the wellbore. In order to differentiate rejuvenated and natural fractures, the reservoir model was separated into two regions. Inner zone includes hydraulic fractures and rejuvenated fractures and outer zone contains natural fractures. Undoubtedly, rejuvenated fracture permeability is higher than natural fracture permeability.

Previous researches successfully proved that iterative coupling between geomechanics and reservoir flow allows easy control of convergence as well as easy maintenance of the reservoir and geomechanics simulators (Tran et al. 2005, 2010). The basic equations of geomechanical model can be decomposed into two sets. One set contains primary flow variables such as pressure and temperature and the other set contains geomechanics variables such as displacement, stress, and strain.

Equilibrium equation, stress-strain relation, and strain-displacement relation for geomechanical effects are given below:

$$\nabla \cdot \boldsymbol{\sigma} - \mathbf{F} = 0 \tag{3.14}$$

$$\boldsymbol{\sigma} = \mathbf{C} : \boldsymbol{\varepsilon} + (\alpha p + \eta \Delta T)\mathbf{I} \tag{3.15}$$

$$\boldsymbol{\varepsilon} = \frac{1}{2}\left[\nabla \mathbf{u} + (\nabla \mathbf{u})^{T}\right] \tag{3.16}$$

where $\boldsymbol{\sigma}$ is the total stress tensor, $\mathbf{F}$ the body force, $\mathbf{C}$ the tangential stiffness tensor, α the Biot's constant, η the thermo-elastic constant, $\boldsymbol{\varepsilon}$ the strain tensor, and $\mathbf{u}$ the displacement vector. Equation 3.14 shows the equilibrium between stress and force in the rock. From combination of these three equations, following equation is obtained:

$$\nabla \cdot \left\{ \mathbf{C} : \frac{1}{2}\left[\nabla \mathbf{u} + (\nabla \mathbf{u})^{T}\right] \right\} = -\nabla \cdot \left[(\alpha p + \eta \nabla T)\mathbf{I}\right] + \mathbf{F} \tag{3.17}$$

The pressure obtained from the primary flow set is used in Eq. 3.17 to solve for the displacement vector. After the displacement vector is determined, the strain and stress tensor can be calculated from Eqs. 3.15 and 3.16, respectively. Then, porosity is computed with these geomechanical factors. Therefore, porosity is not only a function of pressure and temperature but also a function of rock stress and strain.

As mentioned in Sect. 2.6, the deformation of shale reservoir is modeled by stress-dependent correlations coupled with linear-elastic model. Exponential and power-law correlations are used to consider decreasing production caused by porosity and permeability reduction in the shale gas reservoir model. Using effective stress calculated from geomechanical model, porosity and permeability multiplier can be computed. In general reservoirs, assuming Biot's constant to 1, total stress σ is defined as

$$\sigma = \sigma' + p \tag{3.18}$$

Effective stresses of stress-dependent correlations (Eqs. 3.19–3.22) are substituted by Eq. 3.18 and multipliers for porosity and permeability with respect to pressure can be generated.

$$\phi = \phi_{i} e^{-a(\sigma' - \sigma_{i}')} \tag{3.19}$$

$$k = k_{i} e^{-b(\sigma' - \sigma_{i}')} \tag{3.20}$$

$$\phi = \phi_{i} \left(\frac{\sigma'}{\sigma_{i}'}\right)^{-c} \tag{3.21}$$

Fig. 3.1 Schematic view of
shale gas reservoir model

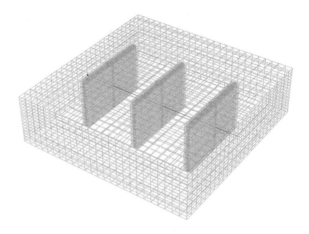

$$k = k_i \left(\frac{\sigma'}{\sigma'_i} \right)^{-d} \tag{3.22}$$

Kim et al. (2015) presented synthetic numerical model with effect of
stress-dependent compaction (Fig. 3.1). The shale gas reservoir with a volume of
$550 \times 550 \times 150$ ft^3 (Size of each grid block is $22 \times 22 \times 30$ ft^3.) is simulated with
no flow outer boundaries. It is assumed that natural fractures exist in every 25 ft in
x and y directions throughout the entire reservoir. The horizontal well is located at
the horizontal and vertical center of the reservoir to produce the gas effectively.
Hydraulic fractures are modeled with a local grid refinement (LGR) technique to
formulate thin blocks assigned with the properties of hydraulic fractures. Within the
LGR, cells increase in size logarithmically away from the fracture and length of the
LGR cells next to the fractures is 1.2 ft. The reservoir is fully penetrated by
hydraulic fractures of which height is the same as net pay. Hydraulic fracture
properties are assumed to be constant along the fracture and have finite conductivity. Fluids include gas and water, but water is residual or immobile so that the
flowing fluid is assumed to be single phase gas. Coefficients in Table 3.1 are applied
in the stress-dependent correlations. Porosity and permeability multipliers based on
laboratory data for both matrix and fractures are shown in Fig. 3.2 Gas
adsorption/desorption and non-Darcy flow are also modeled to depict a real shale
gas reservoir and wellbore storage is neglected. Other basic properties of the
reservoir, hydraulic fracture, and geomechanical model used in the model are listed
in Tables 3.2, 3.3 and 3.4.

Table 3.1 Experimental
coefficients of
stress-dependent porosity and
permeability for shale

a	0.00095 MPa^{-1}
b	0.0353 MPa^{-1}
c	0.033
d	1.478

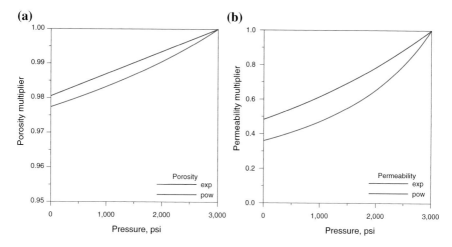

Fig. 3.2 a Porosity and **b** permeability multiplier curves for shale

Table 3.2 Reservoir properties used in the simulation model

Properties	Value
Reservoir pressure (p_i)	3,000 psi
Reservoir temperature (T)	100 °F
Reservoir thickness (h)	150 ft
Matrix porosity (ϕ_m)	0.03
Fracture porosity (ϕ_f)	8.00×10^{-5}
Matrix permeability (k_m)	1.00×10^{-3} md
Natural fracture permeability (k_f)	4.00×10^{-4} md
Gas production rate (q_{sc})	5.00 Mcf/D
Wellbore radius (r_w)	0.25 ft
Horizontal well length (L)	550 ft

Table 3.3 Hydraulic fracture properties used in simulation model

Properties	Value
Hydraulic fracture permeability (k_F)	1,000 md
Fracture half-length (x_F)	100 ft
Fracture height	150 ft
Fracture width	0.001 ft
Fracture spacing (d)	176 ft
Number of fractures (n)	3

Table 3.4 Geomechanical properties used in the simulation model

Properties	Value
Overburden pressure	6,000 psi
Initial effective stress	3,000 psi
Young's modulus	5 GPa
Poisson's ratio	0.2

3.3   Verification with Field Data

In order to verify the generated shale gas model, gas well data from Barnett shale were used to perform history matching. Figure 3.3 shows the daily pressure and gas production data reproduced from Anderson et al. (2010). The horizontal well of the numerical model was drilled 3,250 ft laterally with 19 hydraulic fractures. To apply the exponential and power law correlations to the numerical model of Barnett shale, experimental coefficients should be determined. Cho et al. (2013) presented the coefficient of permeability exponential correlation for this Barnett shale data as 0.0087. Because there is no experimental data for coefficient of power law correlation, it should be calculated by relation between exponential and power law correlations. Table 3.5 reproduced from Dong et al. (2010) presents experimental coefficients determined using curve fitting techniques based on measured permeability and porosity of the tested sandstone and shale samples. From these experimental data, relation functions of exponential and power law correlations could be obtained. Figure 3.4 shows the exponential correlation coefficient vs. power law correlation coefficient plots of porosity and permeability from tested sandstone and shale samples. As shown in Fig. 3.4, all plots show a linear proportional relationship. A power law correlation coefficient of Barnett shale could be obtained from these linear functions. Therefore, using the correlation obtained from Fig. 3.4d, a power law correlation coefficient of permeability was determined as 0.383. In the same way, from Fig. 3.4c, a power law correlation coefficient of porosity was determines as 0.0252. Figure 3.5 provides a porosity and permeability multiplier based on these coefficients for Barnett shale simulation.

Figure 3.6 shows the results of history matching for the well bottomhole pressure in different models. The non-geomechanical model, the geomechanical model with exponential correlation, and the geomechanical model with power law correlation were compared. Models with exponential and power law correlations show lower matching error than a non-geomechanical model. The numerical model that accounts for stress-dependent correlations predicts the production decline of wells in shale gas reservoirs better than a model not considering deformation of rock. In

Fig. 3.3 Daily pressure and gas production data of Barnett shale (reproduced from Anderson et al. 2010)

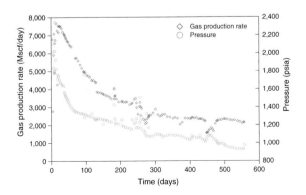

Table 3.5 Experimental coefficients determined using curve fitting techniques based on measured permeability and porosity of the tested sandstone and shale samples (reproduced from Dong et al. 2010)

Sample number	Exponential correlation				Power law correlation			
	Porosity a (MPa^{-1})		Permeability b (MPa^{-1})		Porosity c		Permeability d	
	Loading	Unloading	Loading	Unloading	Loading	Unloading	Loading	Unloading
Sandstone								
R261_sec2_1	0.91×10^{-3}	0.69×10^{-3}	2.84×10^{-3}	1.37×10^{-3}	0.037	0.024	0.120	0.057
R261_sec2_2	1.58×10^{-3}	1.15×10^{-3}	7.68×10^{-3}	2.65×10^{-3}	0.056	0.040	0.303	0.114
R307_sec1	1.03×10^{-3}	0.75×10^{-3}	3.46×10^{-3}	2.16×10^{-3}	0.040	0.028	0.143	0.087
Shale								
R255_sec2_1			16.78×10^{-3}	7.91×10^{-3}			0.844	0.416
R255_sec2_2	0.95×10^{-3}	0.42×10^{-3}	35.29×10^{-3}	18.88×10^{-3}	0.033	0.017	1.478	0.855
R287_sec1	0.94×10^{-3}	0.37×10^{-3}	43.47×10^{-3}	10.58×10^{-3}	0.036	0.016	1.677	0.466
R351_sec2	1.04×10^{-3}	0.82×10^{-3}	25.93×10^{-3}	13.90×10^{-3}	0.032	0.030	0.937	0.514
R316_sec1	1.30×10^{-3}	0.54×10^{-3}			0.046	0.023		
R390_sec3	1.01×10^{-3}	0.65×10^{-3}	42.90×10^{-3}	4.84×10^{-3}	0.036	0.028	1.744	0.196
R437_sec1	0.41×10^{-3}	0.14×10^{-3}	22.78×10^{-3}		0.014	0.006	0.588	

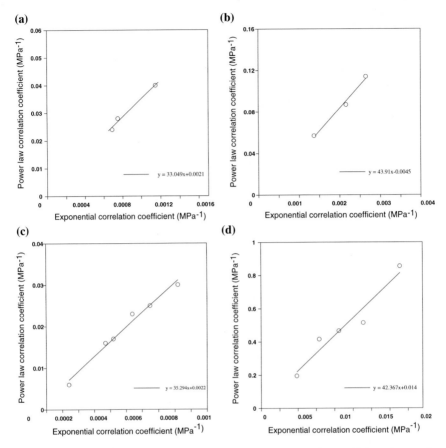

Fig. 3.4 Exponential correlation coefficient versus power law correlation coefficient of **a** porosity and **b** permeability in sand stone and **c** porosity and **d** permeability in shale

addition, matched values of the reservoir properties are also different in all cases. Matched values of fracture and matrix permeability, initial pressure, and hydraulic fracture half-length are presented in Table 3.6, which shows different values caused by stress-dependent compaction effects. Especially, hydraulic fracture half-length of geomechanical models is higher than non-geomechanical model. According to Anderson et al. (2010), results of geomechanical models are more reliable than result of non-geomechanical model.

For the more accurate history matching, concept of SRV is also considered in previous numerical model. Figure 3.7 shows results of history matching considering stress-dependent compaction and SRV. It presents that more exact matching results can be obtained when concept of SRV is considered. In this case, model with exponential correlation and SRV shows more precise result than model with power law correlation and SRV.

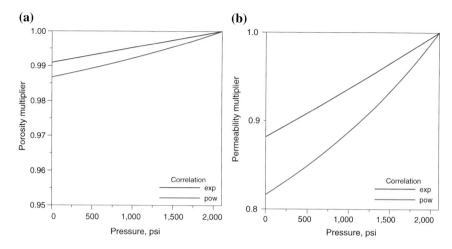

Fig. 3.5 a Porosity and **b** permeability multiplier curves for Barnett Shale model

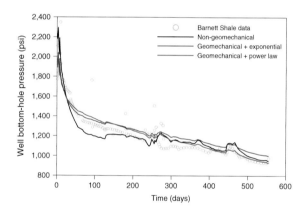

Fig. 3.6 Pressure data and the history matching results of the Barnett shale gas well with non-geomechanical model, geomechanical model with exponential correlation model, and geomechanical model with power law correlation model

Table 3.6 Matched values of fracture and matrix permeability, initial pressure, and hydraulic fracture half-length in non-geomechanical model, geomechanical model with exponential correlation model, and geomechanical model with power law correlation model

Model	Fracture permeability (md)	Matrix permeability (md)	Initial pressure (psi)	Hydraulic fracture half-length (ft)
Non-geomechanical	2.23×10^{-3}	1.00×10^{-7}	2,357	90
Geomechanical + exponential	2.98×10^{-3}	1.78×10^{-6}	2,079	170
Geomechanical + power law	4.71×10^{-3}	4.95×10^{-6}	2,037	174

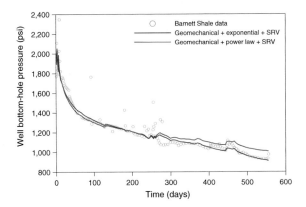

Fig. 3.7 Pressure data and the history matching results of the Barnett shale gas well with geomechanical model with exponential correlation model and power law correlation model when considering SRV concept

References

Anderson D et al (2010) Analysis of production data from fractured shale gas wells. Soc Pet Eng J 15(01):64–75. doi:10.2118/115514-PA

Cho Y et al (2013) Pressure-dependent natural-fracture permeability in shale and its effect on shale-gas well production. SPE Res Eval Eng 16(2):216–228. doi:10.2118/159801-PA

Cipolla CL et al (2010) Reservoir modeling in shale-gas reservoirs. SPE Res Eval Eng 13(4):638–653. doi:10.2118/125530-PA

CMG (2015) GEM user guide. Computer Modelling Group Ltd, Calgary, Alberta

Dong JJ et al (2010) Stress-dependence of the permeability and porosity of sandstone and shale from TCDP Hole-A. Int J Rock Much Min Sci 47(7):1141–1157. doi:10.1016/j.ijrmms.2010.06.019

Evans RD and Civan F (1994) Characterization of non-darcy multiphase flow in petroleum bearing formation. Dissertation, University of Oklahoma

Kazemi H et al (1978) An efficient multicomponent numerical simulator. Soc Pet Eng J 18(05):355–368. doi:10.2118/6890-PA

Kim TH, Lee KS (2015) Pressure-transient characteristics of hydraulically fractured horizontal wells in shale-gas reservoirs with natural- and rejuvenated-fracture networks. J Can Pet Tech 54(04):245–258. doi:10.2118/176027-PA

Kim TH et al (2014) Development and application of type curves for pressure transient analysis of multiple fractured horizontal wells in shale gas reservoirs. Paper presented at the offshore technology conference-Asia, Kuala Lumpur, Malaysia, 25–28 March 2014. doi:10.4043/24881-MS

Kim TH et al (2015) Integrated reservoir flow and geomechanical model to generate type curves for pressure transient responses in shale gas reservoirs. Paper presented at the twenty-fifth international offshore and polar engineering conference, Kona, Hawaii, 21–26 June 2015

Klinkenberg LJ (1941) The permeability of porous media to liquids and gases. In: Drilling and production practice, New York, New York, Jan 1941

Lee SJ et al (2014) Development and application of type curves for pressure transient analysis of horizontal wells in shale gas reservoirs. J Oil Gas Coal T 8(2):117–134. doi:10.1504/IJOGCT.2014.06484

Rubin B (2010) Accurate simulation of non-Darcy flow in stimulated fractured shale reservoirs. Paper presented at the SPE western regional meeting, Anaheim, California, 27–29 May 2010. doi:10.2118/132093-MS

Tran D et al (2005) An overview of iterative coupling between geomechanical deformation and reservoir flow. Paper presented at the SPE international thermal operations and heavy oil symposium, Calgary, Alberta, 1–3 November 2005

Tran D et al (2010) Improved gridding technique for coupling geomechanics to reservoir flow. Soc Pet Eng J 15(1):64–75. doi:10.2118/115514-PA

Warren JE, Root PJ (1963) The behavior of naturally fractured reservoirs. Soc Pet Eng J 3(3):245–255

Yu W et al (2013) Sensitivity analysis of hydraulic fracture geometry in shale gas reservoirs. J Pet Sci Eng 113:1–7. doi:10.1016/j.petrol.2013.12.005

Chapter 4
Performance Analysis

4.1 Introduction

In this chapter, practical approaches for analyzing well performance in shale gas reservoirs are presented. Mini-frac test are performed before main fracturing to determine the reservoir and fracture properties for the main stimulation design. It is generally performed without proppant and the basic principles are analogous to those for pressure analysis of transient fluid flow in the reservoir. Decline curve analysis (DCA), which is used to forecast future production in the conventional reservoirs, can be also used to forecast future production in the shale gas reservoirs. In conventional reservoirs, engineers have used an empirical analysis method, .i.e. exponential and hyperbolic relations, introduced to the industry for evaluating estimated ultimate recovery (EUR). However, these relations are inaccurate in shale reservoirs due to invalid assumptions so that various rate decline relations are proposed recently. Power-law exponential, stretched exponential, Duong, and logistic growth models are presented. There are several rate transient analysis (RTA) methods documented in the literature. Among these, three plots are particularly well suited for tight and shale gas production analysis: square root time plot, flowing material balance plot, and log-log plot. Proper usage of these plots will provide a reliable identification of dominant flow regimes exhibited in the data as well as estimates of bulk reservoir properties, apparent skin and hydrocarbon pore volume (HCPV). Finally, extensive simulations were conducted considering effects of reservoir and fracture properties in the shale reservoir to analyze the pressure behavior in log-log diagnostic plots and productivity index curve. Afterwards, the application of type curve matching technique was provided to determine petrophysical properties of a shale gas reservoir.

© The Author(s) 2016
K.S. Lee and T.H. Kim, *Integrative Understanding of Shale Gas Reservoirs*,
SpringerBriefs in Applied Sciences and Technology,
DOI 10.1007/978-3-319-29296-0_4

4.2 Mini-Frac Test

In the wells to be hydraulically fractured, mini-frac test, also called calibration tests, are frequently performed to determine the parameters needed for the stimulation design (Benelkadi and Tiab 2004). Fracture pressure analysis was pioneered by Nolte (1979, 1988). The basic principles are analogous to those for pressure analysis of transient fluid flow in the reservoir. Both provide a means to interpret complex phenomena occurring underground by analyzing the pressure response resulting from fluid movement in reservoirs.

The analysis of fracturing pressure, during before and after closure, provides a powerful tool for understanding and improving the fracture process. Advances in mini-fracture-analysis techniques have provided methods for the determination of fracturing-treatment design parameters such as leak-off coefficient, fracture dimensions, fluid efficiency, closure pressure, and reservoir parameters. These parameters can then be used to determine the pad volume required, the best fluid-loss additives to be used, and how to achieve the optimum fracturing-treatment design.

Figure 4.1 shows a typical history of the calibration test from the beginning of pumping until the reservoir disturbance. The wellbore is filled with the injection fluid, a pumper is rigged in, and additional fluid is injected to break down the formation and create a short fracture. These are generally performed without proppant. The well is then shut-into observe closure of the fracture and to monitor the after-closure falloff response (Ewens et al. 2012). Pressures during before and after closure periods provide complementary information pertinent to the fracture-design process. The types of analysis have been split into two distinct categories:

1. Before Closure Analysis: After the treatment is complete, the pressure begins to fall off, much like a conventional pressure transient analysis (PTA) of fall-off test. Complications exist because there is a fracture in the media, which is very much in a dynamic state. Over time, this open fracture will close. Before closing the fracture may continue to grow as a result of energy stored during the treatment. The analysis of pressure to closure will give an indication of the fracture closure and leak-off behavior of the combined fracture/porous media. Closure pressure can be determined from pressure decline using the G-function plot, which will be explained later. Furthermore, using G-function time at fracture closure, formation permeability can be estimated by empirical function derived from numerical simulations (Barree et al. 2009).

2. After Closure Analysis: After the fracture is closed, the pressure response loses its dependency on the mechanical response of an open fracture. It is governed by the transient pressure response within the reservoir. This response, which results from fluid loss during fracturing, can exhibit a late time radial response. This flow pattern can be addressed in a manner analogous to conventional well test analysis. In this period, continued analysis of the pressure signature may give information about permeability and other reservoir properties.

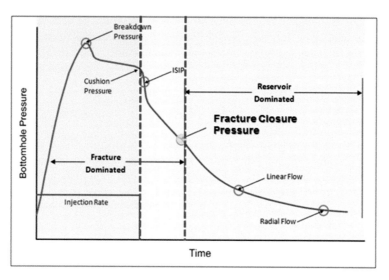

Fig. 4.1 DFIT pressure history (reproduced from Nolte et al. 1988)

4.2.1   Before Closure Analysis

Nolte (1979) introduced a dimensionless function called the G-function, or some-
times just called G time. When the G-function is plotted against the fall-off pressure
after a fracture injection test, a straight line would form under ideal conditions and
the slope would lead to the calculation of the leak-off coefficient (Castillo 1987).
The development of the G-function was based on the assumption that the leak-off
coefficient during fracturing is constant. The basic G-function calculations are based
on the following equations:

$$G(\Delta t_D) = \frac{4}{\pi}\left[g(\Delta t_D) - \frac{4}{3}\right], \tag{4.1}$$

$$g(\Delta t_D) = \frac{4}{3}\left[(1 + \Delta t_D)^{1.5} - \Delta t_D^{1.5}\right], \tag{4.2}$$

$$\Delta t_D = \frac{t - t_p}{t_p} \tag{4.3}$$

where G is the G-function, Δt_D the dimensionless time, g an intermediate variable,
and t_p the time to the end of injection. When the leak-off coefficient is constant, a
plot of pressure versus G should be a straight line. Deviation from such a straight
line would indicate a change in flow regime. Nolte (1986) identified this as the

closure time, from which fracture closure pressure, p_c, could be determined. From the closure time alone, it is then possible to determine the fluid efficiency, η, which is defined as the amount of fluid remaining in the fracture at closure, normalized to the total pumped volume:

$$\eta = \frac{G_c}{2 + G_c},\qquad(4.4)$$

where η is the fluid efficiency and G_c the G-function at closure. This is independent of fracture geometry and is a key parameter in hydraulic fracturing design because it will directly affect the size of the pad. An additional plot used in the industry is the square root time plot, where pressure, p, is plotted against $\sqrt{t}$. It is an empirical plot, based on the general idea of linear flow, as in PTA. However, because the point of initiation of leak-off for fracturing fluid is different for each point in the fracture, the G-function is rigorous, while $\sqrt{t}$ is not.

In practice, picking closure time from a G-function plot has proved to be problematic. A very significant number of wells have pressure-dependent leak-off (PDL) or other nonlinearities that invalidate Nolte's (1986) basic assumptions. Therefore, as with specialized plots in PTA, multiple interpretations are often possible. Approaches and solutions to the PDL problem began with Castillo (1987) and Mukherjee et al. (1991). Barree and Mukherjee (1996) developed a diagnostic plot, which will be called the combination G-function plot, that involves plotting three quantities—p, $\frac{dp}{dG}$, and $G\frac{dp}{dG}$—versus G-function on the same plot. The combination G-function plot is the equivalent of the log-log derivative plot in PTA and can be used to identify flow regimes and to choose p_c.

In the combination G-function plot, the expected signature of the G-function semi-log derivative, $G\frac{dp}{dG}$, is a straight-line through the origin (zero G-function and zero derivative) (Barree 1998). The correct straight line tangent to the semi-log derivative of the pressure versus G-function curve must pass through the origin. Fracture closure is identified by the departure of the semi-log derivative of pressure with respect to G-function from the straight line through the origin. During normal leak-off, with constant fracture surface area and constant permeability, the first derivative ($\frac{dp}{dG}$) should also be constant (Castillo 1987). The primary p versus G curve should follow a straight line (Nolte 1979).

The confirmation of closure pressure also can be done with the square root time plot. The primary p versus $\sqrt{t}$ curve should form a straight line during fracture closure, as with the G-function plot. The indication of closure is the inflection point on the p versus $\sqrt{t}$ plot. However, it is difficult to catch the inflection point so that the best way to find the inflection point is to plot the first derivative of p versus $\sqrt{t}$ and find the point of maximum amplitude of the derivative.

Barree et al. (2009) also found that, for either the constant matrix leak-off case or PDL case, correlation between permeability, k, and G-function at closure, G_c can be presented as follow:

$$k = \frac{0.0086\mu_f\sqrt{0.01p_z}}{\phi c_t \left(\frac{G_c E r_p}{0.038}\right)^{1.96}} \tag{4.5}$$

where μ_f is the mini-frac fluid viscosity (cp), p_z the net fracture extension pressure above closure pressure or $p_z = p_{ISI} - p_c$ (psi), E the Young's modulus (Mpsi), and r_p the storage ratio (dimensionless). The fracture fluid viscosity, μ_f, is normally set to 1.0. The storage ratio, r_p, represents the amount of excess fluid that needs to be leaked off to reach fracture closure when the fracture geometry deviates from the normally assumed constant-height planar fracture. For the constant matrix leak-off and PDL cases, it is 1.0. This permeability correlation allows design work when all other data is lacking.

4.2.2 After Closure Analysis

After fracture closure, transient response is dominant within the reservoir exhibiting linear or radial flow, losing its dependency from the mechanical response of an open fracture. This late time pressure falloff would be a good representation of the reservoir response allowing the estimation of reservoir pressure and permeability. The after closure response is similar to the behavior observed during conventional well test analysis, supporting an analogous methodology for this evaluation.

Gu et al. (1993) have initially developed a method to determine formation permeability using an impulse-fracture test. They derived a solution for mini-frac after closure analysis by considering the leak-off from the fracture as a distribution of instantaneous line sources. Nolte (1997), Nolte et al. (1997), and Talley et al. (1999) have developed several methods for interpreting the after closure period of fracture calibration tests to identify flow regimes and determine reservoir parameters. They developed specialized plots for two special cases after the fracture has closed. The first case assumes the well is in linear flow after closure and the second that the well is in radial flow after closure. For linear flow the corresponding equations are:

$$p(t) - p_i = m_L F_L(t, t_c), \tag{4.6}$$

$$m_L = C_L \sqrt{\frac{\pi\mu}{k\phi c}}, \tag{4.7}$$

$$F_L(t, t_c) = \frac{2}{\pi}\sin^{-1}\left(\sqrt{\frac{t_c}{t}}\right) \quad \text{with } t > t_c, \tag{4.8}$$

where m_L is the slope of the linear flow in after closure analysis plot, F_L the Nolte after closure linear time function, t_c the closure time, and C_L the combined leak-off coefficient. In these formulae, μ refers to the far-field viscosity. For radial flow, they gave the following flow equations:

$$p(t) - p_i = m_R F_R(t, t_c) \tag{4.9}$$

$$m_R = 251{,}000 \left(\frac{\mu V_{inj}}{kht_c} \right), \tag{4.10}$$

$$F_R(t, t_c) = \frac{1}{4} \ln \left(1 + \frac{16}{\pi^2} \frac{t_c}{t - t_c} \right) \quad \text{with } t > t_c, \tag{4.11}$$

where m_R is the slope of the radial flow in after closure analysis plot, F_R the Nolte after closure radial time function, and V_{inj} the injection volume.

Benelkadi and Tiab (2004) proposed modified method for permeability determination by the use of after closure radial analysis. The proposed method is based on the pressure derivative with respect to the radial time function, which is not affected by the value of reservoir pressure. The method is simple because it requires only one log-log plot to identify the radial-flow regime and to determine reservoir parameters. Based on the equations from Gu et al. (1993) and Nolte (1997), the modified method is developed as given below

$$\log(\Delta p) = \log(F_R) + \log(m_R) \tag{4.12}$$

$$\log \left[\frac{d(\Delta p)}{dF_R} \right] = \log(m_R), \tag{4.13}$$

where $\Delta p = p(t) - p_i$.

Equation 4.12 indicate that the radial flow is characterized by a unit slope line, and the intercept with the Δp axis is m_R at $F_R = 1$. With the pressure derivative, from Eq. 4.13, the radial flow is characterized by a horizontal line that intercepts the $\frac{d(\Delta p)}{dF_R}$ axis at m_R. Therefore, the reservoir permeability is determined from m_R. In the log-log plot, only one unit slope line can cross the horizontal line at point m_R of the Δp axis. Thus, to determine the reservoir pressure, the value of the assumed reservoir pressure is varied until the pressure difference curve overlies the drawn unit slope line.

Soliman et al. (2005) used superposition of constant rate solution in Laplace space and performed late time approximations to obtain impulse equations for bilinear, linear and radial flow. Craig and Blasingame (2006) developed an analytical model that accounts for fracture growth, leak-off, closure, and after-closure. The late time approximation of their model produced impulse equations that are the similar with solutions of Soliman et al. (2005). The equations governing bilinear, linear, and radial flow are defined as:

$$p(t) - p_i = 264.6 \frac{V_{inj}}{h} \mu^{\frac{3}{4}} \left(\frac{1}{\phi c_t k}\right)^{\frac{1}{4}} \frac{1}{\sqrt{k_f w_f}} \left(\frac{1}{t_{inj} + \Delta t}\right)^{\frac{3}{4}} \tag{4.14}$$

$$p(t) - p_i = 48.77 \frac{V_{inj}}{h} \sqrt{\frac{\mu}{\phi c_t k x_f^2}} \left(\frac{1}{t_{inj} + \Delta t}\right)^{\frac{1}{2}} \tag{4.15}$$

$$p(t) - p_i = 1694.4 \frac{V_{inj} \mu}{kh} \frac{1}{t_{inj} + \Delta t} \tag{4.16}$$

where k_f the fracture permeability, w_f the fracture width, t_{inj} the injection time, and x_f is the fracture half-length. If the fluid flow regime is radial, permeability can be directly calculated with Eq. 4.16. If the linear flow is shown, the slope of Eq. 4.15 is function of permeability and hydraulic fracture half-length. Hydraulic fracture half-length can be calculated with the permeability from equation of radial flow. In the bilinear flow, the slope of equation is a function of both permeability and fracture conductivity as given in Eq. 4.14. In this case, fracture conductivity also can be determined if the permeability is obtained from radial flow.

4.2.3 Example of Mini-Frac Test

Based on theory mentioned earlier, comprehensive process of mini-frac test with multiple plots was described in this section. Barree (1998) and Barree et al. (2009) described a methodology utilizing plots of G-function, square root of shut-in time, log-log pressure derivative, and Nolte after closure analysis. With these methods, other mentioned techniques were used to ensure a consistent interpretation of the closure process.

Mini-frac test data of gas well A obtained from tight gas reservoir were analyzed. Figures 4.2, 4.3, 4.4, 4.5, 4.6, 4.7, 4.8, 4.9, 4.10 and 4.11 show results of mini-frac test. In before closure analysis, for consistent identification of fracture closure, three techniques are illustrated: G-function, square root of shut-in time, and log-log pressure derivative. All these analyses begin at the instantaneous shut-in pressure (ISIP). The ISIP is taken as the incipient fracture extension pressure. It is defined as final injection pressure minus pressure drop due to friction in the wellbore and perforation of slotted liner.

Figure 4.2 shows G-function plot of gas well A. Constant leak-off is shown by the straight line character of the observed $G \frac{dp}{dG}$ data through the origin. Closure time is picked at deviation from the straight line. In this case, fracture closure pressure, p_c, is 4148 psi, fracture closure time, t_c, is 153.63 min, and G-function at closure,

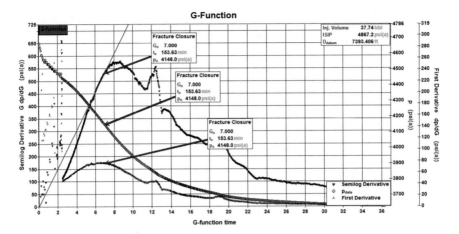

Fig. 4.2 G-function plot for gas well A

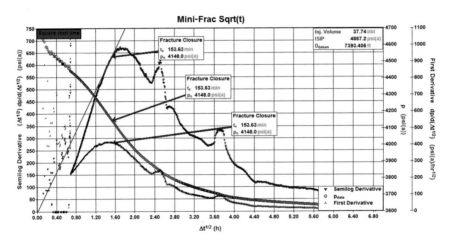

Fig. 4.3 Square root time plot for gas well A

G_C, is 7. These fracture closure values can be confirmed by square root time plot (Fig. 4.3). In this plot, indication of the fracture closure is maximum point of first derivative of p versus $\sqrt{t}$ and also departure from the straight line through the origin on the semi-log derivative of the p versus $\sqrt{t}$ curve. From these G-function and square root time plots, a single closure point can be confirmed. The log-log plot of pressure difference and semi-log derivative is shown in Fig. 4.4. It is common for the pressure difference and semi-log derivative curves to be parallel immediately before closure. In many cases, a near perfect 0.5 slope line is evident, strongly suggesting linear flow in the open fracture. The separation of the two parallel lines always marks fracture closure and is the final confirmation of consistent closure

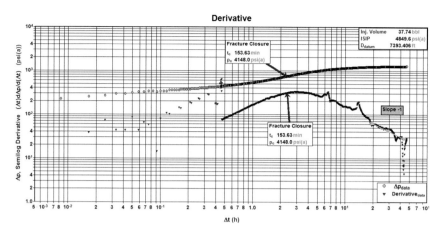

Fig. 4.4 Log-log derivative plot for gas well A

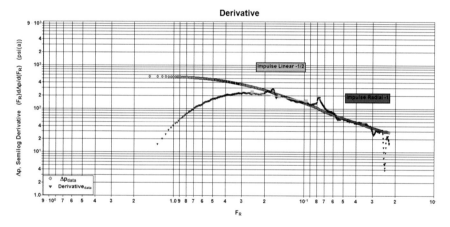

Fig. 4.5 Nolte derivative plot for gas well A

identification. In addition, pressure derivative curve shows flow regime of after closure period. The -1 slope line of the semi-log derivative curve is an indicator of radial flow. If the slope had been -0.5, this would indicate after-closure linear flow.

Figures 4.5, 4.6 and 4.7 show plots of Nolte after closure analysis. For the after closure analysis, identification of flow regime is important. Figure 4.5 shows the plot of semi-log pressure derivative with respect to F_R. In this data, radial and linear flow regimes are identified from the slope of -1 and -0.5 in the semi-long derivative of the pressure curve. From the observed radial flow period, Cartesian radial flow plot can be used to determine permeability and initial reservoir pressure (Fig. 4.6). Using Eq. 4.9, permeability and initial reservoir pressure are estimated as 0.04560 md and 3601 psi. In the same way, using Eq. 4.6 and permeability from

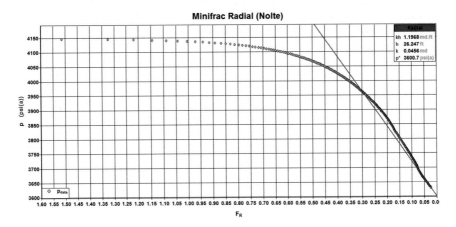

Fig. 4.6 Nolte radial flow plot for gas well A

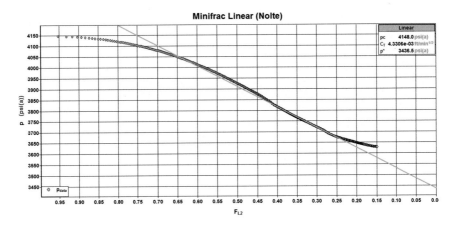

Fig. 4.7 Nolte linear flow plot for gas well A

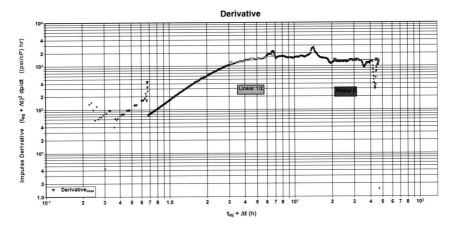

Fig. 4.8 Soliman derivative plot for gas well A

radial flow equation, leak-off coefficient and initial reservoir pressure are calculated as 4.33×10^{-3} ft/min$^{1/2}$ and 3437 psi (Fig. 4.7).

Figures 4.8, 4.9 and 4.10 show plots of Soliman-Craig after closure analysis. Although this technique shows different graphs and equations, the method is similar with Nolte after closure analysis. From the plot of semi-log pressure derivative, radial and linear flow regimes are identified by the slope of 0 and 0.5 (Fig. 4.8). Permeability and initial reservoir pressure are estimated as 0.04262 md and 3598 psi from the radial flow (Fig. 4.9). These values show similar results with Nolte after closure analysis. From linear flow period, initial reservoir is calculated as 3334 psi (Fig. 4.10). In this period, fracture half-length also can be calculated if

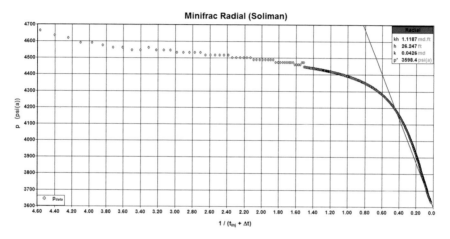

Fig. 4.9 Soliman radial flow plot for gas well A

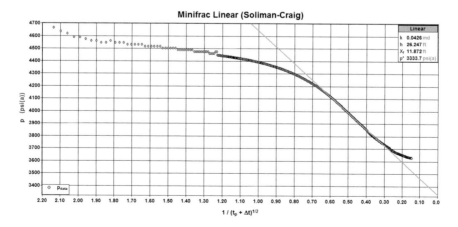

Fig. 4.10 Soliman linear flow plot for gas well A

permeability was determined from radial flow. In this case, fracture half-length is 11.9 ft.

Figure 4.11 shows after closure analysis of Benelkadi and Tiab (2004). Permeability can be determined from horizontal line of pressure derivative as 0.04865 md. By matching the intercept of unit-slope line of pressure difference with that of horizontal line of pressure derivative, initial reservoir pressure can be determined. From this method, initial reservoir pressure is estimated as 3603 psi. Results from after closure analysis are compared in Table 4.1. Estimated initial pressure is separated in radial and linear flow regimes. Results show the consistency of after closure analysis.

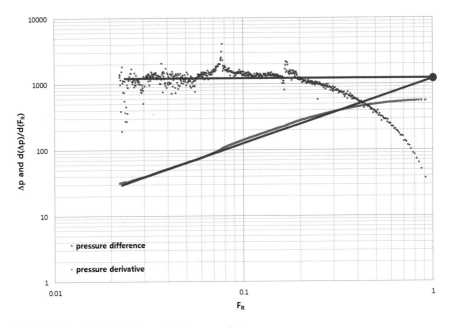

Fig. 4.11 Benelkadi radial flow plot for gas well A

Table 4.1 Results of reservoir permeability and initial pressure estimation using mini-frac test

	Nolte	Soliman-Craig	Benelkadi and Tiab
Permeability (md)	0.04560	0.04262	0.04865
Pressure: radial flow (psi)	3601	3598	3603
Pressure: linear flow (psi)	3437	3334	

4.3 Decline Curve Analysis

Decline curve analysis (DCA) is one of the most common techniques used to forecast future production performance of conventional reservoirs. In the last decade, DCA has been also used to forecast individual well performance in unconventional reservoirs. However, the application of DCA in unconventional plays could be problematic. A prerequisite to any discussion on DCA for unconventional plays is the understanding that no simplified time-rate model can accurately capture all elements of performance. From a historical perspective, DCA and production forecasting using Arps' exponential and hyperbolic relations have been the standard for evaluating estimated ultimate recovery (EUR) in petroleum engineering (Houze et al. 2015). However, in unconventional plays such as shale gas, tight/shale oil reservoirs, these relations often yield ambiguous results due to invalid assumptions. The main assumptions which form the basis of traditional DCA can be summarized as:

1. There is no significant change in operating conditions and field development during the producing life of the well.
2. The well is producing with a constant bottomhole flowing pressure.
3. There is a boundary-dominated flow regime and reservoir depletion was established.

In ultra-low permeability reservoir systems, it is common to observe basic violations of the assumptions related to traditional DCA. Hence, the misapplications of the Arps' relations to production data often result in significant overestimation of reserves, specifically when the hyperbolic relation is extrapolated with a b-exponent greater than one. In order to prevent overestimation of EUR, a hyperbolic trend may be coupled to an exponential decline at late time. However, this approach remains empirical and may be 'non-unique' in the hands of most users, yielding widely varying estimates of reserves.

The issues with Arps' relations have led numerous authors to propose various rate decline relations: power-law exponential (Ilk et al. 2008), stretched exponential (Valko 2009), Duong (2011), and logistic growth model (Clark et al. 2011) which attempt to model the time-rate behavior observed in unconventional plays. Specifically, these relations focused on characterizing the early time transient and transitional flow behavior. They are based on empirical observations of characteristic behaviors of certain plays. None of them are sufficient to forecast production for all unconventional plays. In other words, one equation may work for one play and perform poorly on another one. It is, therefore, important to understand the behavior of each equation, and apply these relations appropriately for production forecasts.

4.3.1   The Arps Equations

Arps' hyperbolic relations (1945) are widely used in DCA for production extrap-
olations and reserves estimations. The basis for the Arps' relations is empirical.
Johnson and Bollens (1927) and later Arps (1945) presented the decline parameter,
loss-ratio and derivative of the loss-ratio functions as:

$$D(t) \equiv -\frac{1}{q(t)}\frac{dq(t)}{dt} \tag{4.17}$$

$$\frac{1}{D(t)} \equiv -\frac{q(t)}{\frac{dq(t)}{dt}} \tag{4.18}$$

$$b(t) \equiv \frac{d}{dt}\left[\frac{1}{D(t)}\right] \equiv -\frac{d}{dt}\left[\frac{q(t)}{\frac{dq(t)}{dt}}\right] \tag{4.19}$$

Equations 4.17 and 4.18 are empirical results based on observations. For the case
of $D =$ constant, Eq. 4.17 does yield the exponential decline which can be derived
for the case of pseudosteady state (or boundary-dominated) flow in a closed
reservoir containing a constant compressibility liquid and being produced at a
constant wellbore flowing pressure. The exponential rate decline relation is given
as:

$$q(t) = q_i \exp(-D_i t) \tag{4.20}$$

where D_i is the Arps' initial decline rate of hyperbolic model. Ilk et al. (2008)
provide and alternate computation of the D and b-parameter using rate-cumulative
data. The alternate D-parameter formulation is given as:

$$D(t) \equiv -\frac{dq(t)}{dQ(t)} \tag{4.21}$$

The alternate b-parameter formulation is given by:

$$b(t) \equiv q\frac{d}{dQ(t)}\left[\frac{1}{D(t)}\right] \tag{4.22}$$

For reference, Blasingame and Rushing (2005) provide the derivation of the
"hyperbolic" rate decline relation in complete detail. They defined D-parameter for
a hyperbolic rate decline given as:

$$D(t) \equiv \frac{1}{\frac{1}{D_i} + bt} \tag{4.23}$$

Completing the derivation (Blasingame and Rushing 2005), the hyperbolic rate decline relation is given as:

$$q(t) = \frac{q_i}{(1 + bD_i t)^{\frac{1}{b}}} \qquad (4.24)$$

It is possible to infer exponential or hyperbolic behavior by observing the D and b-parameters. A constant D-parameter indicates exponential decline. A constant b-parameter indicates hyperbolic decline. For matching purposes, the user should first adjust b from the b-parameter plot, and then match the D-parameter with a model. The initial rate (q_i) can be adjusted to complete the match and obtain the production forecast. It is possible to use a segmented hyperbolic if the user identifies multiple constant trends of the b-value.

Finally, it is important to note that industry wide application of the Arps' hyperbolic relation in unconventional reservoirs includes a modification with the exponential decline at later times to prevent overestimation of reserves as the hyperbolic equation is unbounded for b-values greater than one (i.e. transient flow assumption for b-values greater than one). Hyperbolic decline is switched to an exponential decline once a certain yearly decline value is reached. This yearly decline value is set by the analyst and is called as the 'terminal decline' value. This protocol yields the 'modified' hyperbolic designation. Figure 4.12 shows decline curve analysis of a single well example with Arps decline curve in Marcellus shale.

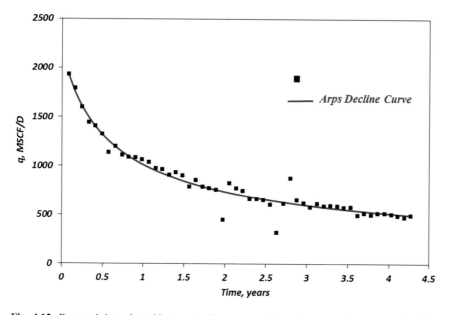

Fig. 4.12 Rate and time plot with Arps decline curve in Marcellus shale (Nelson et al. 2014)

4.3.2 Power Law Exponential Model

The power law exponential relation was derived by Ilk et al. (2008) exclusively from the observed behavior of the D-parameter and b-parameter. Its primary assumption is that the D-parameter exhibits a straight line behavior on a log-log scale, which essentially corresponds to a power-law model. The resulting differential equation yields the power-law exponential relation if the D-parameter formulation is approximated to a power law model as follow:

$$D = D_\infty + D_1 t^{-(1-n)}. \tag{4.25}$$

where n is the exponent, D_∞ the decline parameter at infinite time $(t = \infty)$, and D_1 the decline parameter intercept at day 1 $(t = 1)$. By introducing a constraining variable (D_∞), the loss ratio can be approximated by a decaying power law function with a constant behavior at large times in contrast to the hyperbolic relation. This variable converts the power-law exponential equation to an exponential decline with a smooth transition. However, in almost all of the applications in unconventional reservoirs, D_∞ is not required since there has been no observation of the constant D-parameter trend and the nature of the power-law exponential relation is conservative as it models the b-parameter trend declining with time. Power-law exponential relation is obtained by substituting Eq. 4.25 into Eq. 4.17 given below:

$$q(t) = \hat{q}_i \exp\left(-\hat{D}_i t^n - D_\infty t\right) \tag{4.26}$$

The application of the power-law exponential relation is centered on the use of the D-parameter and time plot. Once the straight line is identified, slope and intercept values associated with the $\hat{D}_i$ and n parameters are obtained. The $\hat{q}_i$ parameter is adjusted to achieve the match on rate and time plot.

Schematic plot of the hyperbolic and power law exponential models is described in Fig. 4.13 (Ilk et al. 2008). In Fig. 4.13, for the hyperbolic relation, the D-parameter has a near-constant behavior at early times and a unit-slope, power law decay at late times. As one may expect, for the "power law loss ratio" relation the D-parameter exhibits a power law decay behavior from transient through transition flow, and then turns gently towards a constant value (i.e., D_∞) at very large times.

4.3.3 Stretched Exponential Production Decline Model

The stretched exponential relation is essentially the same as the power-law exponential relation without the constraining variable (D_∞). Outside petroleum engineering, the stretched exponential relation has many applications such as in physics where numerous processes manifest this behavior. In geophysics, the stretched exponential function is used to model aftershock decay rates.

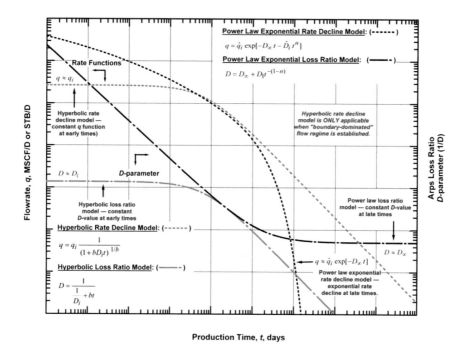

Fig. 4.13 Rate decline and loss ratio curves of hyperbolic and power law exponential models (Ilk et al. 2008)

In general, the stretched exponential function is used to represent decays in randomly disordered, chaotic, heterogeneous systems. It can be suggested that the stretched exponential decay of a quantity is generated by a sum (superposition) of exponential decays with various time constants. This leads to the interpretation of heterogeneity where production decline in an unconventional reservoir system is determined by a great number of contributing individual volumes exhibiting exponential decays with a specific distribution of time constants. As the Arps' original decline curve model, the stretched exponential model is completely empirical. In contrast to Arps' method, however, this model is based on a differential equation. The differential equation of the model and stretched exponential function is given as (Valko 2009; Valko and Lee 2010):

$$\frac{dq(t)}{dt} = -n\left(\frac{t}{\tau_{SEPD}}\right)^n \frac{q}{t} \tag{4.27}$$

$$q(t) = \hat{q}_i \exp\left[-\left(\frac{t}{\tau_{SEPD}}\right)^n\right] \tag{4.28}$$

where τ_{SEPD} the characteristic time parameter (in simple terms the analogue to the concept of half-life). Although Valko (2009) did not try to develop a "rate-time" analysis relation at first, he utilized the form given by 4.28 as a means of evaluating a database of production data. The stretched exponential relation can be applied in the same manner as the power-law exponential relation using diagnostic plots or alternatively the procedure described by Valko (2009) could be applied. Figure 4.14 describes the application of the stretched exponential production decline model on a specific field example.

4.3.4   The Duong Model

Most of the production data from shale reservoirs exhibit fracture-dominated flow regimes and rarely reach late-time flow regimes. This indicates that traditional approaches for production decline do not work in shale reservoirs. Duong (2011) proposed new approach of production decline in which fracture flow is dominant and matrix contribution is negligible.

If a fracture flow regime is prolonged over the life of a well, the gas flow rate q will be:

$$q(t) = q_1 t^{-n}, \tag{4.29}$$

where q_1 is the flow rate at day 1 and n one-half or one-quarter for linear flow or bilinear flow. The gas cumulative G_p will be

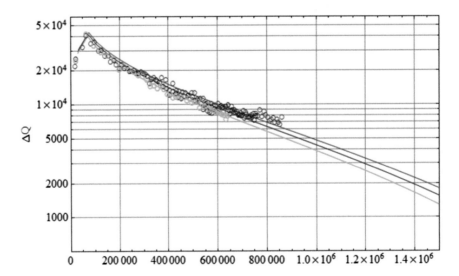

Fig. 4.14 Fitted decline curves for the average wells (Valko and Lee 2010)

$$G_p(t) = \int\limits_0^t q\,dt = q_1 \frac{t^{1-n}}{1-n}. \tag{4.30}$$

From Eqs. 4.29 and 4.30,

$$\frac{q(t)}{G_p(t)} = \frac{1-n}{t}. \tag{4.31}$$

A log-log plot of rate over cumulative production versus time will yield a straight line with a unity slope regardless of fracture types with ideal assumption. In practice, a slope of greater than unity is normally observed because of actual field operations, data approximation, and flow-regime changes. Log-log plots for rate over cumulative production versus time from field data give a straight line with a negative slope, $-m_{Dng}$, and an intercept of a_{Dng} given below:

$$\frac{q(t)}{G_p(t)} = a_{Dng}t^{-m_{Dng}}. \tag{4.32}$$

From above equation, equations for q and G_p were derived from Duong (2011) as follow:

$$q(t) = q_1 t^{-m_{Dng}} \exp\left[\frac{a_{Dng}}{1-m_{Dng}}\left(t^{1-m_{Dng}} - 1\right)\right] = t\left(a_{Dng}, m_{Dng}\right) \tag{4.33}$$

$$G_p(t) = \frac{q_1}{a_{Dng}} \exp\left[\frac{a_{Dng}}{1-m_{Dng}}\left(t^{1-m_{Dng}} - 1\right)\right] \tag{4.34}$$

Duong (2011) suggested a step-by-step procedure of how to perform decline analysis with Duong model (Fig. 4.15). First, the production data histories are plotted and checked. Then, the log-log plot of rate over cumulative production versus time is constructed to determine the $-m_{Dng}$ and a_{Dng} with Eq. 4.32. After determining these values, gas flow rate is plotted against $t\left(a_{Dng}, m_{Dng}\right)$ to obtain q_1 with Eq. 4.33. Other models, such as the power-law exponential, the stretched exponential and the logistic growth, account for deviations at later times. Such deviations also occur when a terminal decline is imposed on the modified-hyperbolic relation. Therefore, the EUR estimates from Duong's model are higher unless a constraining variable is also imposed. The linear flow assumption of the Duong model may hold for some plays, but it will generally need modifications to deal with changes in flow regimes (i.e. transitional flow, depletion of SRV, interference, etc.).

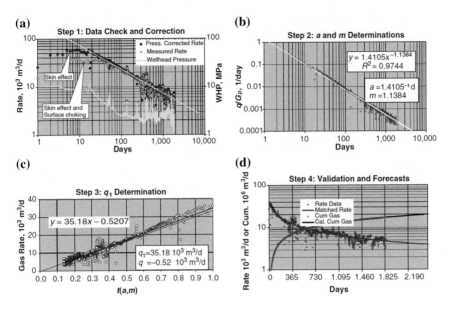

Fig. 4.15 Four steps for using the Duong model (2011)

4.3.5 Logistic Growth Model

Logistic growth curves are a family of mathematical models used to forecast growth in numerous applications. Conceptually, logistic growth models assume that the growth variable increases then stabilizes. Logistic growth models have a term called the carrying capacity, which is the size at which the growth variable stabilizes and growth rate terminates. Clark et al. (2011) utilizes the logistic growth model for forecasting cumulative production of the wells in unconventional oil and gas reservoirs. The logistic growth model to describe cumulative production and rate is given below:

$$G_p(t) = \frac{Kt^{n_{LGM}}}{a_{LGM} + t^{n_{LGM}}} \tag{4.35}$$

$$q(t) = \frac{dG_p(t)}{dt} = \frac{Kn_{LGM}a_{LGM}t^{n_{LGM}-1}}{\left(a_{LGM} + t^{n_{LGM}}\right)^2} \tag{4.36}$$

The parameter K is the carrying capacity and referred to as the ultimate of oil and gas recovery from the well without any economic limits. This parameter is included in the model itself. Cumulative production will approach K while the rate tends to zero. The parameter n_{LGM} controls the decline. When n_{LGM} tends to one, the decline becomes steeper. The parameter a_{LGM} controls the time at which half of the carrying capacity is reached. A high value of a_{LGM} indicates stable production. A low value

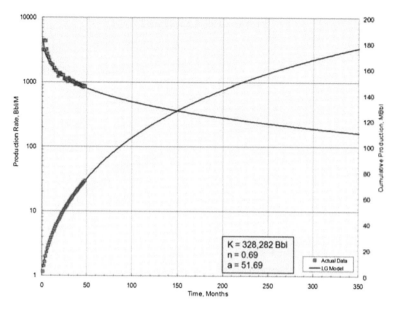

Fig. 4.16 Data from example Bakken Shale well fit with logistic growth model (Clark et al. 2011)

of a_{LGM} points to a steeper decline. Figure 4.16 shows an example of both rate and cumulative versus time production data for a well in Bakken Shale being fit with the logistic growth model.

4.3.6 Conclusions

DCA is a fast and efficient but empirical way to forecast production into the future under certain assumptions (Houze et al. 2015). All of the equations may produce good matches across the entire production and a EUR value can be estimated associated with each model. However, Fig. 4.17 presents an example where all decline curve relations [Arps, PLE (power law exponential), SEPD (stretched exponential decline), Duong, and LGM (logistic growth model)] match the entire production data and differences are observed at late times due to specific model behavior. As mentioned earlier, none of these relations have a direct link to reservoir engineering theory other than analogy. At this point, one must assume that each of these models can be considered as empirical in nature and generally center on a particular flow regime and/or characteristic data behavior. A useful way to apply decline curve analysis is to apply all equations together to obtain a range of results rather than a single EUR value. This range of results may be associated with the uncertainty related to the production forecast and can be evaluated as a function of time.

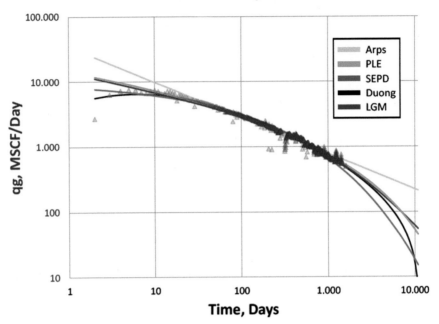

Fig. 4.17 Rate and time plot matched with five decline curve models (Kanfar and Wattenbarger 2012)

It is very optimistic to suggest that decline curve relations may approximate to, or match model-based (time-pressure-rate) analysis profiles. These relations cannot capture all elements of the complexity of fluid flow behavior in unconventional reservoirs modelled by reservoir solutions (analytically or numerically). However, the average trend can be used to approximate the behavior. Certain flow regimes can be approximated with a constant b value in the hyperbolic model. Along these lines decline curve relations may also be used as proxies to represent model-based analysis (i.e. time-rate-pressure analysis) forecasts in economic software.

4.4 Rate Transient Analysis

Conventional production analysis assumes constant flowing bottomhole pressure, drainage area, permeability, skin, and existence of boundary dominated flow. Most of these assumptions are no longer valid in unconventional reservoirs. Therefore, it is crucial that not only rate, but also pressure and other reservoir parameters are taken into account to properly evaluate unconventional wells and determine the true flow capacity of their reservoir in linear transient flow (Belyadi et al. 2015). The ultra-low permeability matrix of shale reservoir provides stable long-term

production. Rate transient analysis methods for analyzing production data are well
documented in the literature (Anderson 2010). Three plots are particularly well
suited for tight and shale gas production analysis: square root time plot, flowing
material balance plot, and log-log plot. Proper usage of these plots will provide a
reliable identification of dominant flow regimes exhibited in the data as well as
estimates of bulk reservoir properties such as $A\sqrt{k}$, apparent skin and hydrocarbon
pore volume (HCPV). Armed with this information, then suitable reservoir models
can be constructed to generate type curves and forecast a long-term production for
estimating reserves.

4.4.1   Square Root Time Plot

The square root-time plot, $\frac{m(p_i)-m(p_{wf})}{q}$ versus $\sqrt{t}$, is probably the single most
important plot for characterizing long-term shale gas well performance (Fig. 4.18).
This is because fractured shale gas reservoirs will typically be dominated by linear
flow. Linear flow appears as a straight line on the square root-time plot as follow:

$$\frac{m(p_i) - m(p_{wf})}{q} = m_{sqr}\sqrt{t} \tag{4.37}$$

where m_{sqr} is the slope of straight line during linear flow period in square root time
plot. In some cases, the observed linear flow may prevail for several years. It is
assumed that the observed linear flow is a result of transient matrix drainage into the

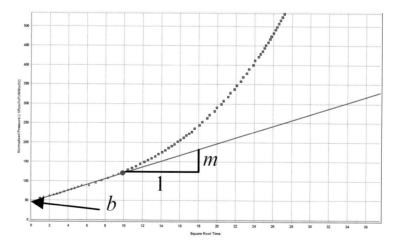

Fig. 4.18 Square root time plot (Anderson et al. 2010)

fractures. This is a reasonable assumption, but it may not be the case if the fracture spacing is very dense and/or conductivity is low.

The slope of the square root time plot yields the linear flow parameter (LFP), which is the product of flow area and square root of permeability:

$$LFP = A\sqrt{k} = \frac{630.8T}{m_{sqr}} \frac{1}{\sqrt{(\phi\mu_g c_t)_i}} \qquad (4.38)$$

where A is the half of total matrix surface area draining into fracture system. There is no way to decouple the flow area from the permeability in linear flow analysis. One must be independently estimated before the other can be determined. It should be noted that Eq. 4.38 is derived based on the assumption of a constant flowing pressure at the well. The constant pressure solution is assumed as many shale gas wells produce under high drawdown due to the extremely low reservoir permeability.

Consider a single vertical fracture of length, x, as shown in Fig. 4.19a. The A in $A\sqrt{k}$ would now be defined as the product of the fracture length, x, and the net pay thickness, h. Equation 4.38 can be used to calculate the permeability as follows:

$$k = \left(\frac{LFP}{xh}\right)^2 \qquad (4.39)$$

If a cased horizontal well with multiple parallel fractures are equally spaced, as shown in Fig. 4.19b, then the area becomes the sum of all the individual fracture areas.

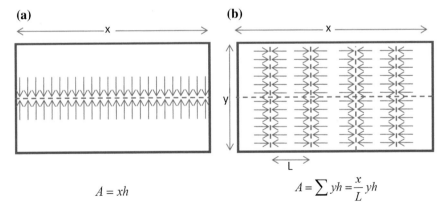

(a)

$A = xh$

(b)

$A = \sum yh = \frac{x}{L} yh$

Fig. 4.19 Illustration of linear flow in a fractured reservoir. **a** Single fracture. **b** Multiple transverse fractures (Anderson et al. 2010)

$$A = \sum yh = \frac{x}{L}yh = \frac{A_{SRV}}{L}h \qquad (4.40)$$

In the equation, x is the horizontal well length, y the stimulated reservoir width, A_{SRV} the area of the SRV, and L the fracture spacing. Following equation is obtained by combining Eqs. 4.38 and 4.40:

$$k = \left(\frac{LFP \times L}{xyh}\right)^2 = \left(\frac{LFP \times L}{A_{SRV}h}\right)^2 \qquad (4.41)$$

There are three unknowns in Eq. 4.41: k (permeability), L (fracture spacing) and y (stimulated reservoir width). Thus, two of these should be independently specified. As presented in the following section, stimulated reservoir width can be estimated from the interpreted SRV on the Flowing Material Balance (FMB) plot, provided that boundary-dominated flow is achieved. In the absence of boundary-dominated flow, a suitable stimulated reservoir width is chosen based on microseismic (if available), well spacing or analogs. As stated previously, the range of expected permeability for shales is from 1 to 100 nd. Thus, upon choosing suitable matrix permeability, fracture spacing can be calculated by Eq. 4.42.

$$L = \frac{xyh\sqrt{k}}{LFP} \qquad (4.42)$$

In fractured shale gas wells exhibiting matrix to fracture linear flow, a significant skin effect can be observed from the pressure loss due to finite conductivity in the fracture system, even if there is no mechanical skin damage at the wellbore. This skin effect may have a significant impact on well productivity and therefore, is an important parameter for production forecasting. The y-intercept on the square root-time plot, b, represents a constant pressure loss, from which the apparent skin, s', can be calculated using the following equation:

$$s' = \frac{kh}{1417T}b \qquad (4.43)$$

4.4.2   Flowing Material Balance Plot

The mode of boundary-dominated flow seen in conventional reservoirs results from the pressure transient investigating all of the surrounding no-flow boundaries in the system. The boundaries may be natural features such as faults or pinchouts, or in multi-well reservoirs, simply the borders between drainage areas of adjacent wells. It is unlikely that this mechanism would be observed in fractured shale gas reservoirs, as the matrix permeability is too low to enable investigation of large areas (Anderson et al. 2010). However, apparent boundary-dominated flow is seen in some shale gas production data sets. This is not true boundary-dominated flow, but rather depletion of the matrix blocks resulting from interference between adjacent fractures within the stimulated reservoir volume. Figure 4.20b illustrates the expected layout of no-flow boundaries (caused by interference) for two different fracture system geometries, and compares them against the conventional reservoir shown in Fig. 4.20a. This apparent boundary-dominated flow should, in theory, be followed by infinite acting flow as the unstimulated matrix surrounding the SRV continues to contribute to the production response.

Mattar and McNeil (1998) proposed a flowing gas material balance method for constant rate case without shut-into compute gas-in-place. Their analysis is based on the fact the pressure at any point in the reservoir declines at the same rate during constant rate boundary-dominated flow. Thus, the pressure drop measured at the wellbore is the same as the pressure drop that would be observed anywhere in the reservoir for constant rate boundary-dominated flow. Consequently, the authors shifted the straight line depicted by a plot of sandface or wellhead flowing pressure versus cumulative production to the initial reservoir or initial wellhead pressure to yield gas-in-place on the x-intercept. Mattar and Anderson (2003) proposed a flowing material balance method based on the modified version of Agarwal-Gardener rate/cumulative type curves. Their analysis involves a plot of pseudopressure drop normalized rate against pseudopressure drop normalized cumulative on a linear scale. Their analysis yields initial-fluid-in-place on the x-intercept. The authors defined their normalized cumulative in terms of material balance pseudotime.

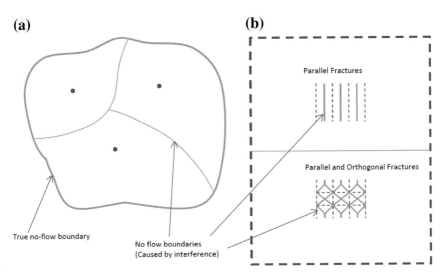

Fig. 4.20 Boundary dominated flow in **a** conventional reservoir versus **b** fractured shale reservoir (Anderson et al. 2010)

The Flowing Material Balance (FMB) is a production data analysis method, based on a modified version of the Agarwal-Gardner rate-cumulative type curves (Matter and Anderson 2003). The method is similar to a conventional material balance analysis, but requires no shut-in pressure data (except initial reservoir pressure). Instead, it uses the concepts of pressure normalized rate and material balance (pseudo) time to create a simple linear plot, which extrapolates to fluids-in-place.

From the pseudosteady state equation for gas reservoir, using pseudopressure and pseudotime:

$$\frac{\Delta m(p)}{q} = \frac{2p_i}{(\mu_g c_t Z)_i G_i} t_{ca} + b'_{pss} \tag{4.44}$$

where

$$b'_{pss} = \frac{1.417 \times 10^6 T}{kh} \left(\ln \frac{r_e}{r_{wa}} - \frac{3}{4} \right)$$

In Eq. 4.44, G_i is the original gas in place, t_{ca} the material balance pseudotime, and b'_{pss} the y-intercept of normalized PSS equation for gas (also called inverse productivity index). Multiplying both sides of Eq. 4.44 by $\frac{q}{\Delta m(p)}$, dividing by b_{pss}, and rearranging, we get:

$$\frac{q}{\Delta m(p)} = -\frac{2qt_{ca}p_i}{(\mu_g c_t Z)_i \Delta m(p)} \frac{1}{G_i b'_{pss}} + \frac{1}{b'_{pss}} \tag{4.45}$$

In Eq. 4.45, a plot of normalized rate, $\frac{q}{\Delta m(p)}$, versus normalized cumulative rate, $\frac{2qt_{ca}p_i}{(\mu_g c_t Z)_i \Delta m(p)}$, yields a straight line with an x-intercept of initial gas-in-place (G_i). This FMB analysis plot can be used to determine the connected hydrocarbon pore volume (HCPV) from a boundary-dominated flow signal, which will appear as a straight line on the graph (Fig. 4.21). This boundary-dominated flow signal is representative of the SRV. If boundary-dominated flow is not exhibited on the log-log plot, the FMB analysis plot should not be used to determine the SRV. In this case, an independent interpretation of the SRV would be required.

4.4.3 Log-Log Diagnostic Plot

It is difficult to maintain a constant bottomhole pressure during production due to the ever-changing operating conditions. Palacio and Blasingame (1993) introduced the material balance time function that enables us to analyze variable rate/pressure

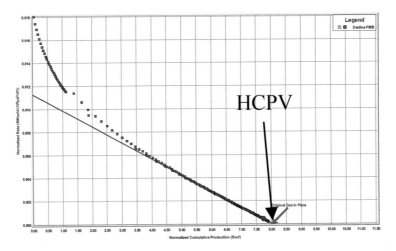

Fig. 4.21 Flowing material balance plot (Anderson et al. 2010)

data. The material balance time is defined as the ratio of cumulative production to instantaneous rate:

$$t_c = \frac{Q}{q} \tag{4.46}$$

where t_c is the material balance time, Q the cumulative production, q the flow rate. Material balance pseudotime, which accounts for changing gas compressibility and viscosity, is used in case of gas. Material balance pseudotime is defined as follows:

$$t_{ca} = \frac{(\mu_g c_t)_i}{q_g} \int_0^t \frac{q_g}{\bar{\mu}_g \bar{c}_t} dt \tag{4.47}$$

where $(\mu_g)_i$ is the gas viscosity at initial reservoir condition, $(c_t)_i$ the total compressibility at initial reservoir condition, q_g the gas flow rate, $\bar{\mu}_g$ the gas viscosity at the average reservoir pressure, $\bar{c}_t$ the total compressibility at the average reservoir pressure. According to Agarwal et al. (1999), constant rate and constant bottomhole pressure cases show identical results when material balance time is used. In other words, material balance time ensures that the constant pressure solution can be adapted for the constant rate solution which is widely used in pressure transient analysis.

Palacio and Blasingame (1993) and Doublet et al. (1994) presented pressure normalized rate that is defined as the rate divided by pressure drop. Pressure normalized rate and derivative with respect to material balance time are computed as

$$\frac{q}{\Delta m} \tag{4.48}$$

$$\left(\frac{q}{\Delta m}\right)_d = \frac{d\left(\frac{q}{\Delta m}\right)}{d \ln t_{ca}} \tag{4.49}$$

where

$$\Delta m = m(p_i) - m(p_{wf})$$

Equation 4.50 shows the definition of the normalized rate integral that is the cumulative average of the normalized rate when plotted against material balance time. Using integral function, any noises in the raw data are effectively removed whereupon a smooth decline curve is obtained. The integral curve is similar to the original production decline curve, but is far smoother. The rate integral and integral-derivative functions are defined as

$$\left(\frac{q}{\Delta m}\right)_i = \frac{\int_0^{t_{ca}} \frac{q}{\Delta m} dt_{ca}}{t_{ca}} \tag{4.50}$$

$$\left(\frac{q}{\Delta m}\right)_{id} = \frac{d\left(\frac{q}{\Delta m}\right)_i}{d \ln t_{ca}} \tag{4.51}$$

Rate normalized pressure is the inverse of pressure normalized rate, so that these functions are essentially same. It is of significant importance because rate normalized pressure functions have distinct characteristic features from pressure normalized rate. Rate normalized pressure, derivative, integral, and integral-derivative functions with respect to material balance pseudotime are given below.

$$\frac{\Delta m}{q} \tag{4.52}$$

$$\left(\frac{\Delta m}{q}\right)_d = \frac{d\left(\frac{\Delta m}{q}\right)}{d \ln t_{ca}} \tag{4.53}$$

$$\left(\frac{\Delta m}{q}\right)_i = \frac{\int_0^{t_{ca}} \frac{\Delta m}{q} dt_{ca}}{t_{ca}} \tag{4.54}$$

$$\left(\frac{\Delta m}{q}\right)_{id} = \frac{d\left(\frac{\Delta m}{q}\right)_i}{d \ln t_{ca}} \tag{4.55}$$

To identify flow regimes in shale gas reservoir, normalized rate/pseudopressure data, derivative, integral, and integral-derivative functions were used along with material balance pseudotime. Even if rate and pressure functions do not show substantive difference in analyzing data, using both formats is encouraged (Ilk et al. 2010). Application of both formats which have distinct characteristic features ensures that relevant diagnostic results are derived from the production data.

4.5 Reservoir Performance Evaluation

Due to the ultra-low matrix permeability and complex natural fractures of shale gas reservoirs, it would show a considerably long period and intricate flow regimes in transient flow periods. Therefore, understanding pressure behavior of hydraulically fractured horizontal well is of importance to provide a perception into a long-term production performance as well as to present criteria for the estimate of reservoir and fracture parameters with pressure transient analysis. In this section, extensive simulations were conducted considering effects of reservoir and fracture properties in the shale reservoir. The main purpose of this work is to analyze the pressure behavior in log-log diagnostic plots and productivity index curve. Afterwards, the application of type curve matching technique was provided to determine petro-physical properties of a shale gas reservoir. Results from this study provide insights into the pressure transient characteristics and estimation of reservoir properties during production from shale gas reservoir through a multi-fractured horizontal well.

4.5.1 Pressure Transient Characteristics

Because shale gas reservoirs have extremely low matrix permeability, transient flow regimes take a long time to produce gas from a well. Therefore, pressure transient characteristics for shale gas wells are of importance and have been discussed in several studies. The pressure transient behavior of a hydraulically fractured horizontal well was studied by Larsen and Hegre (1991, 1994). They described pressure transient flow regimes with corresponding analytical solutions. Horne and Temeng (1995) developed an analytical model to describe the inflow performance and transient pressure behavior of a horizontal well with multiple hydraulic fractures. The effects of the number, position, and direction of fractures on pressure transient responses were discussed by Raghavan et al. (1997) in high-permeability conventional reservoirs. Mederios et al. (2007) presented a discussion of diagnostic pressure and pressure derivative plots for hydraulically fractured horizontal wells in locally and globally fractured formations. They compared the performances of horizontal wells with longitudinal and transverse fractures. Medeiros et al. (2008) explored the influence of matrix permeability, fracture spacing, and well spacing on pressure behavior in tight gas reservoirs. Lu et al. (2009) explained the pressure behavior of horizontal wells in dual-porosity, dual-permeability naturally fractured reservoirs. Cheng (2011) researched the pressure transient behaviors of a horizontal well with hydraulic fractures using a numerical simulation model with consideration of various factors in a range essentially practical to Marcellus Shale. Wu et al. (2012) studied a numerical model for modeling transient gas flow behavior and its application to well testing analysis for a hydraulically fractured vertical well in unconventional gas reservoirs. Rana and Ertekin (2012) presented a new set of type curves for pressure transient analysis of composite dual porosity systems. The composite, double porosity system represented shale gas reservoirs with multistage hydraulically fractured horizontal wells. Lee et al. (2014) addressed pressure transient analysis for horizontal wells in shale gas reservoirs incorporating a number of important formation properties and nonlinear processes. They provided various type curves in terms of dimensionless pseudopressure and time for transient pressure responses and conducted type curve matching for synthetic pressure data. Kim et al. (2014) presented a comprehensive reservoir simulation model to investigate the characteristics of pressure transient responses under the influences of hydraulic fracture properties and nonlinear gas flow mechanisms. Results from those numerical simulations showed various flow periods in log-log plots of pseudopressure and derivatives of pseudopressure versus time.

Typically, shale gas reservoirs consist of a matrix and natural fractures. Hydraulic fracturing not only creates new fractures but also rejuvenates existing natural fractures, which opens networks of interconnected fractures around the wellbore. Kim and Lee (2015) considered the distinction between natural fractures affected by hydraulic fracturing and those that are unaffected. The following sections introduce their new model, called SRV model, based on real shale gas reservoirs, including rejuvenated fractures and natural fractures.

Figure 4.22 is a log-log plot of the SRV model with pseudopressure and pseudopressure derivative versus time (Kim and Lee 2015). In accordance with the pseudopressure derivative curve, which is a more effective way to analyze flow regimes than the pseudopressure curve, flow regimes are identified below, and distributions of pressure drops are shown in Fig. 4.23.

- The first flat straight line from point A to B represents the fracture radial flow (FRF). FRF occurs mainly in hydraulic fractures.
- The next period, from point B to C, has a convex shape for the downward direction associated with dual porosity systems. As mentioned above, the difference in fracture pressure and matrix pressure is increasing in the early part of this period and then decreasing. At the end of this period, the fracture and matrix pressures achieve dynamic equilibrium.
- The section from point C to D is the transient flow period displaying various flow regimes. In this model, bilinear flow (BLF) and inner linear flow (ILF) are observed.
- From point D to E is a transition period affected by the inner region, including natural and rejuvenated fractures, and the outer region containing only natural fractures. Because of the high contrast of permeability between these two regions, their boundary behaves like a leaking boundary. In this paper, this

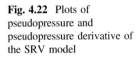

Fig. 4.22 Plots of pseudopressure and pseudopressure derivative of the SRV model

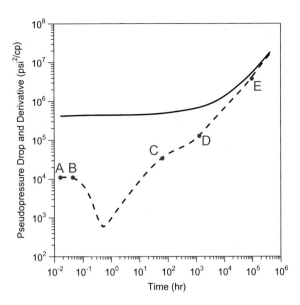

transition flow is defined as pseudo-boundary dominated flow (PBDF) and shows a slope between 1/2 and 1.

– The final log-log straight line with unit slope after point E represents the boundary dominated flow (BDF) period. If this period is observed, it means that the pressure of both the matrix blocks and the fractures has encountered external boundaries.

All the flow regimes in this list can be switched or not be revealed, depending on specific conditions, as described below.

Figure 4.24 shows the pseudopressure and pseudopressure derivative curves of the SRV model and the existing model used in previous researches for comparison. The existing model considers only hydraulic fractures and natural fractures and does not reflect the effect of the rejuvenated fractures considered in SRV model. To compare the difference in effects in the outer region including only natural fractures, the rejuvenated fracture permeability of the SRV model is equal to the natural fracture permeability of the existing model. In the pseudopressure derivative curves of Fig. 4.24, the difference in the two models appears after point D. As defined earlier, the SRV model displays PBDF. In the existing model, however, transition and a compound linear flow (CLF, Fig. 4.23e) period are observed. Because the permeability of fractures is constant in the whole existing model, CLF appears perpendicular to the horizontal well. The beginning of BDF is also different between the two models. Because pressure propagation is impeded by PBDF in the SRV model, the time to reach BDF in the existing model is 8000 h shorter than that of the SRV model.

The effects of fracture and matrix permeability both inside and outside of SRV, hydraulic fracture properties, and range of SRV are discussed using log-log plots of

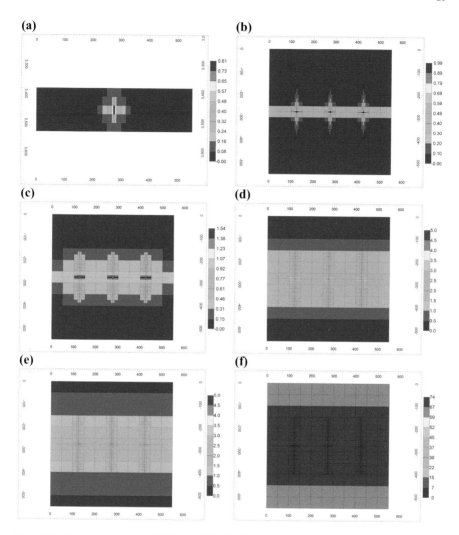

Fig. 4.23 Flow regimes of the SRV model (**a** FRF, **b** BLF, **c** ILF, **d** PBDF, **f** BDF) and existing model (**a** FRF, **b** BLF, **c** ILF, **e** CLF, **f** BDF)

pseudopressure and pseudopressure derivative (Kim and Lee 2015). Inner fracture permeability, or rejuvenated fracture permeability, affects flow regimes throughout the production life except for the late time. As rejuvenated fracture permeability increases, the internal flow of the hydraulic fractures rapidly arrives at equilibrium so that different flow regimes appear in each case after the dual porosity flow period. Inner matrix permeability affects the period of dual porosity flow. Large inner matrix permeability enhances matrix flow toward the rejuvenated fractures and advances pressure equilibrium. Outer fracture permeability, or natural fracture

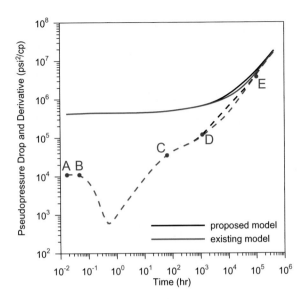

Fig. 4.24 Comparison of the SRV model and the existing model on pseudopressure and pseudopressure derivative

permeability, affects the slope of PBDF and the beginning of BDF on derivative curves. Hydraulic fracture width, height, and half-length affect early, intermediate, and late productivity, respectively. As the range of SRV increases, the slope of PBDF also rises, and BDF begins sooner.

Several papers studied decreasing fracture conductivity depending on reservoir stress. Pedrosa (1986) presented the permeability modulus, which measures the exponential dependency of permeability on pressure to construct type curves from stress-sensitive reservoirs. Tran et al. (2005) proposed several methods for coupling geomechanics to fluid flow in the reservoir. Raghavan and Chin (2004) showed a productivity reduction with three stress-dependent permeability correlations in an isotropic, linear-elastic model. Dong et al. (2010) measured the stress-dependent porosity and permeability. They showed that the data can be fitted by using an exponential correlation and a power law correlation. Cho et al. (2013) presented the effect of pressure-dependent natural fracture permeability with experiments for Bakken-shale core samples and a history matching process.

To analyze the effects of the stress-dependent compaction, Kim et al. (2015) considered five base cases. Two of them are a non-geomechanical model and a non-geomechanical model with pressure dependent exponential correlation. Non-geomechanical model does not consider any geomechanical effects so that porosity changes slightly based on rock compressibility but permeability does not changes. In non-geomechanical model with exponential correlation, porosity and permeability change with pressure dependent exponential correlations. The others are geomechanical model, geomechanical model with exponential correlation, and geomechanical model with power law correlation. In geomechanical model, reservoir properties are computed by iterative coupling between geomechanics and

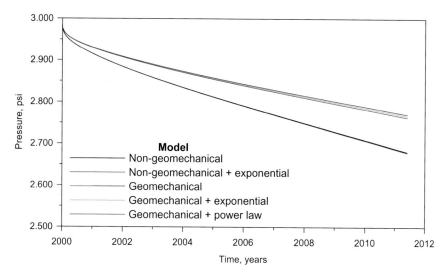

Fig. 4.25   Pressure versus time curves for five base cases

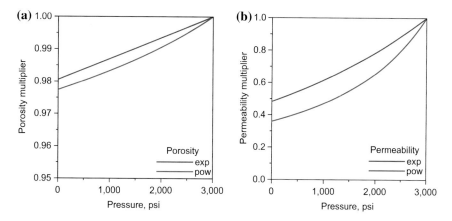

Fig. 4.26   a Porosity and **b** permeability multiplier curves for shale

reservoir flow as mentioned earlier but permeability does not changes. Geomechanical models with exponential and power law correlation consider stress dependent porosity and permeability with geomechanical effects.

A pressure versus time plot for the five base cases is presented in Fig. 4.25 to indirectly analyze productivity. The geomechanical model plays an important role according to this plot. An increase in productivity due to the geomechanical effect is roughly 3–5 %. Deformation of the shale reservoir decreases the pore volume of the matrix and fracture so that the production is activated. The effects of stress-dependent correlations are small compared with the geomechanical model.

Productivities decrease from 1 to 3 % with these correlations due to a reduction in porosity and permeability. Because the reduction in the porosity and permeability multiplier is higher in the power law correlation than the exponential correlation (Fig. 4.26), the reduction in productivity is higher in the model employing the power law correlation than the model with the exponential correlation.

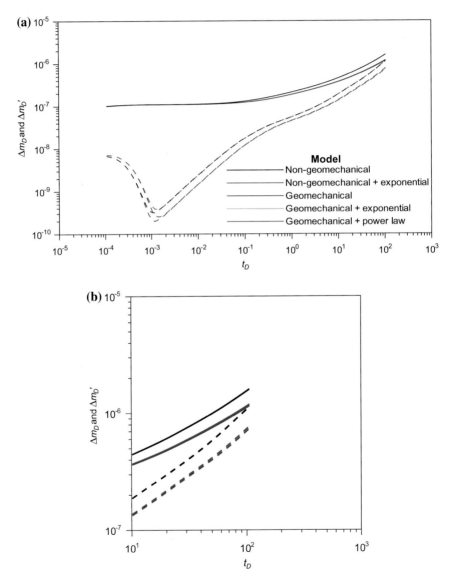

Fig. 4.27 Type curves of dimensionless pseudopressure drop (*solid line*) and derivative (*dashed line*) for five base cases in **a** all periods and **b** later time

To analyze the pressure response specifically, type curves in log-log plots of dimensionless pseudopressure and the pseudopressure derivative as a function of dimensionless time are presented in Fig. 4.27. In all cases, fracture radial flow (FRF), bilinear flow (BLF), and inner linear flow (ILF) are observed (Fig. 4.28). Due to the effect of the geomechanical model, the dual porosity flow period, presenting a downward convex shape, is extended. The dual porosity flow period is the process that the difference between fracture and matrix pressure achieves dynamic equilibrium. Owing to the deformation of the reservoir, equilibrium time is prolonged in the geomechanical model. Stress-dependent correlations tend to extend FRF in early time and increase the slope of ILF in later times (Fig. 4.27b).

Experimental coefficients for stress-dependent correlations, initial effective stress, initial reservoir pressure, natural fracture permeability, matrix porosity, Young's modulus, and Poisson's ratio affect the flow regimes of the shale gas reservoir (Kim et al. 2015). This is particularly apparent in the results of permeability and porosity; geomechanical effects are of importance for low permeability and low porosity reservoirs. Therefore, considering geomechanical effects in a shale gas reservoir is necessary.

Figure 4.29 presents the effect of initial effective stress on porosity and the permeability multiplier. Considerable differences among exponential and power law correlations are observed in these curves. As shown in Fig. 4.29, only the power law correlation is influenced by initial effective stress, while the exponential correlation is not. In the power law correlation, the reduction of porosity and permeability becomes higher as initial effective stress becomes lower. This is because the model with low initial effective stress is more deformable than a model with high initial effective stress.

In order to verify the proposed models mentioned earlier, gas well data from Barnett Shale were used. Figure 3.3 shows the daily pressure and gas production data reproduced from Anderson et al. (2010). Because field data includes variable pressure/rate and noise, corrections were made using material balance time function introduced by Palacio and Blasingame (1993) and rate normalized pressure to analyze variable pressure/rate data. Due to ambient noise of field data, it is difficult

(a)   **(b)**   **(c)**

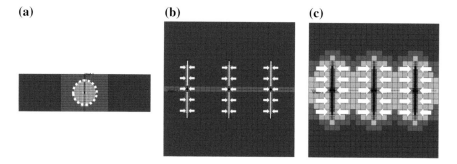

Fig. 4.28 Flow regimes for multi-fractured horizontal shale gas well **a** FRF, **b** BLF, and **c** ILF

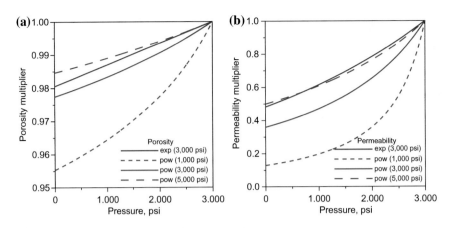

Fig. 4.29 a Porosity and **b** permeability multiplier curves for different initial stress

to analyze slope of flow regimes from derivative curves. Any noise in the raw data is effectively removed when integral functions which are introduced in Sect. 4.4.3.

Figure 4.30 shows rate normalized pressure integral and integral derivative curves for Barnett Shale data. In this plot, three flow regimes mentioned earlier are

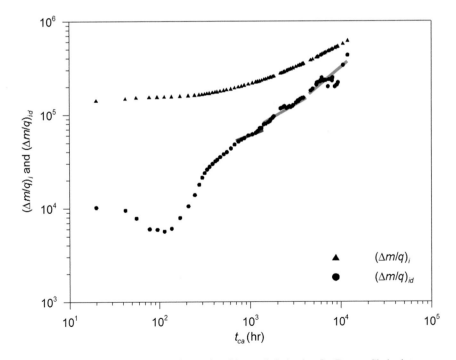

Fig. 4.30 Rate normalized pressure integral and integral-derivative for Barnett Shale data

observed. From 700 to 1500 h, BLF with slope of 0.25 is shown. ILF are observed between 1500 and 4000 h with slope of 0.5. After ILF, slope of derivative curve are roughly 0.8. This period is PBDF which is affected by difference of properties between SRV and outer region.

To verify the effects described in this paper, the field production data of a Marcellus Shale is used. Figure 4.31 shows the flowing bottomhole pressure and gas production rate reproduced from Yeager and Meyer (2010). In the same way with Barnett Shale example, material balance time function and rate normalized pressure with integral functions are used to reduce the effects of variable pressure/rate and noise of field data.

Rate normalized pressure integral and normalized pressure integral derivative curves for Marcellus Shale data is shown in Fig. 4.32. In this plot, three flow regimes mentioned earlier are identified. BLF with slope of 0.25 is shown from 100 to 250 h. From 250 to 500 h, ILF is observed with slope of 0.5. After ILF with brief transition period, slope of derivative curve is measured 0.5 again. This period is CLF. Although all flow regimes mentioned in this paper are not obtained due to quality and quantity of data, the field example presents the verification for the proposed numerical model of shale gas reservoir.

4.5.2 Productivity Index

The transient productivity index is a convenient means of discussing the productivity of horizontal wells in tight formations (Medeiros et al. 2008; Ozkan et al. 2011).

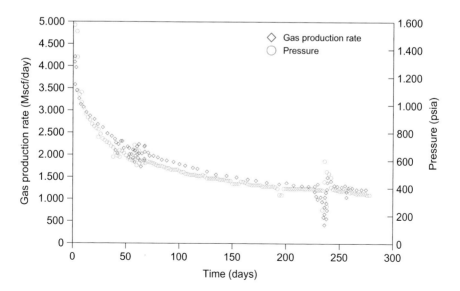

Fig. 4.31 Daily pressure and gas production data of Barnett shale (reproduced from Yeager and Meyer 2010)

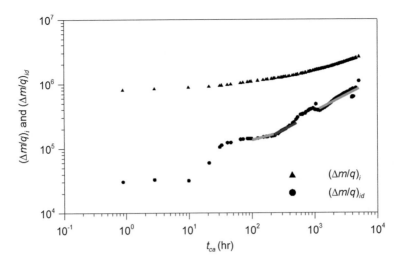

Fig. 4.32 Rate normalized pressure integral and integral-derivative for Marcellus Shale data

The use of a generalized transient productivity index for declining type curve analysis of horizontal wells has been discussed by Araya and Ozkan (2002). The transient productivity index J (scf · psi^2/D · cp) for gas flow is defined as a function of gas pseudopressure as follows:

$$J(t) = \frac{q_{sc}(t)}{\Delta m(p_{wf}) - \Delta m(p_{avg})} \tag{4.56}$$

In compliance with Eq. 4.56, productivity versus a time curve is represented in Fig. 4.33. In the productivity index curve, two flat slope periods are displayed in the early and late parts of the plot. The first flat period is associated with dual porosity systems, corresponding to the period from point B to C of the pseudopressure derivative curve in Fig. 4.22. According with the intermediate straight line of nearly a zero slope in Fig. 2.2, the productivity index is also constant during this period. The straight line with a slope of zero at the end of the productivity index curve indicates the BDF period. Under the semi-steady state condition indicating BDF, the general form of the inflow equation (Dietz 1965) is given by:

$$m(p_{avg}) - m(p_{wf}) = \frac{q\mu}{2\pi kh}\left(\frac{1}{2}\ln\frac{4A}{\gamma C_A r_w^2}\right) \tag{4.57}$$

where A is the area being drained and C_A the Dietz shape factor. From Eqs. 4.56 and 4.57, BDF represents constant productivity and correspondingly displays a flat period at the end of the transient productivity index plot (Fig. 4.33).

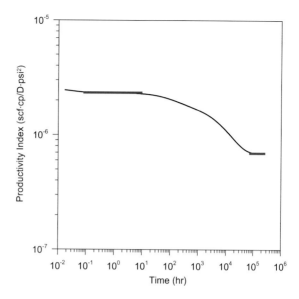

Fig. 4.33 Productivity plot of the SRV model

4.5.3   Type Curve Matching

Based on the extensive numerical simulations with the updated model, a series of type curves were developed in terms of dimensionless pseudopressure drop and derivative versus dimensionless time with respect to various reservoir and fracture properties (Kim et al. 2015; Kim and Lee 2015). Using the developed type curve sets, a simple and practical procedure was presented to estimate reservoir properties in multi-fractured horizontal wells. A step-by-step procedure for analyzing pressure transient tests using these type curves is presented.

Kim and Lee (2015) generated type curves based on pressure data and dimensionless variables. Dimensionless variables for generating type curves with hydraulically fractured horizontal well in shale gas reservoir are presented by Nobakht et al. (2012). The dimensionless terms of time t_D, pseudopressure drop Δm_D, interporosity flow coefficient λ, fracture conductivity F_{CD}, and fracture height h_{DF} are defined as

$$t_D = \frac{0.00633 k_f t}{\phi_{m+f}(c_t \mu)_i x_F^2} \tag{4.58}$$

$$\Delta m_D = \frac{k_f h}{1.417 \times 10^6 T} \frac{\Delta m}{q} \tag{4.59}$$

$$\lambda = \alpha r_w^2 \frac{k_m}{k_f} \tag{4.60}$$

$$F_{CD} = \frac{k_F w_F}{k_m x_F} \tag{4.61}$$

$$h_{DF} = \frac{h_F}{x_F} \tag{4.62}$$

where k_f is the natural fracture permeability, ϕ_{m+f} the porosity of matrix and fractures, x_f the hydraulic fracture half-length, h the reservoir height, T the reservoir
temperature, α the parameter characteristic of the system geometry, k_F the hydraulic
fracture permeability, k_m the matrix permeability, w_F the hydraulic fracture width, and
h_F the hydraulic fracture height. Type curves with dimensionless interporosity flow,
fracture conductivity, and fracture height are presented in Figs. 4.34, 4.35 and 4.36.

Generated type curves present the possibility of predicting reservoir properties
with type curve matching technique when enough data on reservoir and fracture
properties are available. Pressure drawdown data from numerical simulation are
used to predict fracture permeability and total porosity. Other known properties and
procedure for type curve matching for case study are listed below.

p_i = 1500 psi	T = 100 °F	c_t = 5.849 × 10^{-4} psi^{-1}
μ = 0.014 cp	h = 150 ft	x_F = 100 ft

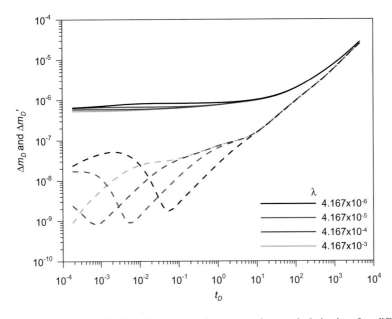

Fig. 4.34 Type curves of dimensionless pseudopressure drop and derivative for different
interporosity flow coefficients

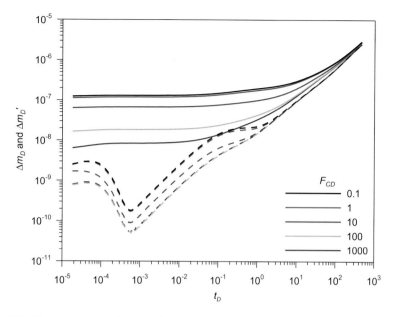

Fig. 4.35 Type curves of dimensionless pseudopressure drop and derivative for different dimensionless fracture conductivities

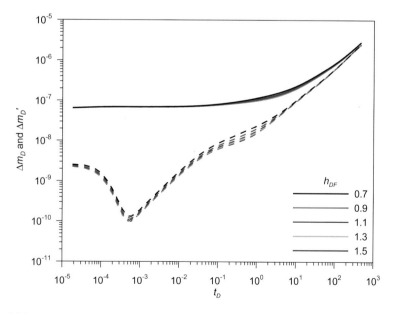

Fig. 4.36 Type curves of dimensionless pseudopressure drop and derivative for different dimensionless fracture heights

Step-1   Calculate $\Delta m(p)$ and $t\Delta m'(p)$.
Step-2   Plot ($\Delta m(p)$ vs. t) and ($t\Delta m'(p)$ vs. t) on a log-log plot.
Step-3   Find the best matching type curve with the log-log plot as shown in Fig. 4.37.
Step-4   Read from any match point.

$$t_M = 10^{-2}, t_{DM} = 1.1 \times 10^{-4}, \Delta m_M = 10^2, \Delta m_{DM} = 1.5 \times 10^{-10}$$

Step-5   Calculate fracture permeability (k_f) using Eq. 4.59.

$$k_f = \frac{1.417 \times 10^6 \times 560 \times 500}{150} \frac{1.5 \times 10^{-10}}{10^2} = 3.97 \times 10^{-3} \, \text{md}$$

Step-6   Calculate total porosity (ϕ_{m+f}) using Eq. 4.58.

$$\phi_{m+f} = \frac{0.00633 \times 3.97 \times 10^{-3}}{5.849 \times 10^{-4} \times 0.014 \times 100^2} \frac{100}{1.1 \times 10^{-4}} = 2.92 \times 10^{-2}$$

The actual values of fracture permeability and total porosity used to generate synthetic test data are 4×10^{-3} md and 3.008×10^{-2}, respectively. Therefore, the calculated reservoir properties from type curve matching are reliable.

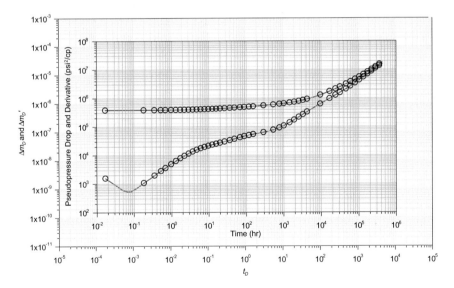

Fig. 4.37 Type curve matching plot for case study

References

Agarwal RG et al (1999) Analyzing well production data using combined-type-curve and decline-curve analysis concepts. SPE Res Eval Eng 2(5):478–486. doi:10.2118/57916-PA

Anderson D et al (2010) Analysis of production data from fractured shale gas wells. Soc Pet Eng J 15(01):64–75. doi:10.2118/115514-PA

Araya A, Ozkan E (2002) An account of decline-type-curve analysis of vertical, fractured, and horizontal well production data. Paper presented at SPE annual technical conference and exhibition, San Antonio, Texas, 29 Sept–2 Oct 2002. doi:10.2118/77690-MS

Arps JJ (1945) Analysis of decline curves. Trans AIME 160(1):228–247. doi:10.2118/945228-G

Barree RD (1998) Applications of pre-frac injection/falloff tests in fissured reservoirs—field examples, SPE Rocky mountain regional. Paper presented at the low-permeability reservoirs symposium, Denver, Colorado, 5–8 April 1998. doi:10.2118/39932-MS

Barree RD, Mukherjee H (1996) Determination of pressure dependent leakoff and its effect on fracture geometry. Paper presented at the SPE annual technical conference and exhibition, Denver, Colorado, 6–9 Oct 1996. doi:10.2118/36424-MS

Barree RD et al (2009) Holistic fracture diagnostics: consistent interpretation of prefrac injection tests using multiple analysis methods. SPE Prod Oper 24(3):396–406. doi:10.2118/107877-PA

Belyadi H et al (2015) Production analysis using rate transient analysis. Paper presented at the SPE eastern regional meeting, Morgantown, West Virginia, 13–15 Oct. doi:10.2118/177293-MS

Benelkadi S, Tiab D (2004) Reservoir permeability determination using after-closure period analysis of calibration tests. Paper presented at the SPE Permian basin oil and gas recovery conference, Midland, Texas, 15–17 May 2001. doi:10.2118/70062-MS

Blasingame TA, Rushing JA (2005) A production-based method for direct estimation of gas-in-place and reserves. Paper presented at the 2005 SPE eastern regional meeting, Morgantown, West Virginia, 14–16 Sept 2005. doi:10.2118/98042-MS

Castillo JL (1987) Modified fracture pressure decline analysis including pressure-dependent leakoff. Paper presented at the low permeability reservoirs symposium, Denver, Colorado, 18–19 May 1987. doi:10.2118/16417-MS

Cheng Y (2011) Pressure transient characteristics of hydraulically fractured horizontal shale gas wells. Paper presented at the SPE eastern regional meeting, Columbus, Ohio, 17–19 Aug 2011. doi:10.2118/149311-MS

Cho Y et al (2013) Pressure-dependent natural-fracture permeability in shale and its effect on shale-gas well production. SPE Res Eval Eng 16(2):216–228. doi:10.2118/159801-PA

Clark AJ et al (2011) Production forecasting with logistic growth models. Paper presented at the SPE annual technical conference and exhibition, Denver, Colorado, 30 Oct–2 Nov 2011. doi:10.2118/144790-MS

Craig DP, Blasingame TA (2006) Application of a new fracture-injection/falloff model accounting for propagating, dilated, and closing hydraulic fractures. Paper presented at the SPE gas technology symposium, Calgary, Alberta, Canada, 15–17 May 2006. doi:10.2118/100578-MS

Dietz DN (1965) Determination of average reservoir pressure from build-up surveys. J Pet Tech 17(8):955–959. doi:10.2118/1156-PA

Dong JJ et al (2010) Stress-dependence of the permeability and porosity of sandstone and shale from TCDP Hole-A. Int J Rock Much Min Sci 47(7):1141–1157. doi:10.1016/j.ijrmms.2010.06.019

Doublet LE et al (1994) Decline-curve analysis using type curves—analysis of oil well production data using material balance time: application to field cases. Paper presented at the international petroleum conference and exhibition of Mexico, Veracruz, Mexico, 10–13 Oct 1994. doi:10.2118/28688-MS

Duong AN (2011) Rate-decline analysis for fracture-dominated shale reservoirs. SPE Res Eval Eng 14(3):377–387. doi:10.2118/137748-PA

Ewens et al (2012) Executing minifrac tests and interpreting after-closure data for determining reservoir characteristics in unconventional reservoirs. Paper presented at the SPE Canadian unconventional resources conference, Calgary, Alberta, Canada, 30 Oct–1 Nov. doi:10.2118/162779-MS

Gu H et al (1993) Formation permeability determination using impulse-fracture injection. Paper presented at the production operations symposium, Oklahoma City, Oklahoma, 21–23 March 1993. doi:10.2118/25425-MS

Horne RN, Temeng KO (1995) Relative productivities and pressure transient modeling of horizontal wells with multiple fractures. Paper presented at the SPE middle east oil show, Bahrain, 11–14 March 1995. doi:10.2118/29891-MS

Houze O et al (2015) Dynamic Data Analysis. KAPPA, Paris

Ilk D et al (2008) Exponential vs. hyperbolic decline in tight gas sands—understanding the origin and implications for reserve estimates using Arps' decline curves. Paper presented at the SPE annual technical conference and exhibition, Denver, Colorado, 21–24 Sept 2008. doi:10.2118/116731-MS

Ilk D et al (2010) Production data analysis—challenges, pitfalls, diagnostics. SPE Res Eval Eng 13 (3):538–552. doi:10.2118/102048-PA

Johnson RH, Bollens AL (1927) The loss ratio method of extrapolating oil well decline curves. Trans AIME 77(01):771–778. doi:10.2118/927771-G

Kanfar MS, Wattenbarger RA (2012) Comparison of empirical decline curve methods for shale wells. Paper presented at the SPE Canadian unconventional resources conference, Calgary, Alberta, Canada, 30 Oct–1 Nov 2012. doi:10.2118/162648-MS

Kim TH, Lee KS (2015) Pressure-transient characteristics of hydraulically fractured horizontal wells in shale-gas reservoirs with natural- and rejuvenated-fracture networks. J Can Pet Tech 54(04):245–258. doi:10.2118/176027-PA

Kim TH et al (2014) Development and application of type curves for pressure transient analysis of multiple fractured horizontal wells in shale gas reservoirs. Paper presented at the offshore technology conference-Asia, Kuala Lumpur, Malaysia, 25–28 March 2014. doi: 10.4043/24881-MS

Kim TH et al (2015) Integrated reservoir flow and geomechanical model to generate type curves for pressure transient responses in shale gas reservoirs. Paper presented at the twenty-fifth international offshore and polar engineering conference, Kona, Hawaii, 21–26 June 2015

Larsen L, Hegre TM (1991) Pressure transient behavior of horizontal wells with finite-conductivity vertical fractures. Paper presented at the international Arctic technology conference, Anchorage, Alaska 29–31 May 1991

Larsen L, Hegre TM (1994) Pressure transient analysis of multifractured horizontal wells. Paper presented at the SPE annual technical conference and exhibition, New Orleans, Louisiana, 25–28 Sept 1994. doi:10.2118/28389-MS

Lee SJ et al (2014) Development and application of type curves for pressure transient analysis of horizontal wells in shale gas reservoirs. J Oil Gas Coal T 8(2):117–134. doi:10.1504/IJOGCT.2014.06484

Lu J et al (2009) Pressure behavior of horizontal wells in dual-porosity, dual-permeability naturally fractured reservoirs. Paper presented at the SPE middle east oil and gas show and conference, Bahrain, 15–18 March 2009. doi:10.2118/120103-MS

Mattar L, Anderson DM (2003) A systematic and comprehensive methodology for advanced analysis of production data. Paper presented at the SPE annual technical conference and exhibition, Denver, Colorado, 5–8 Oct 2003. doi:10.2118/84472-MS

Mattar L, McNeil R (1998) The "flowing" gas material balance. J Can Pet Tech 37(02):52–55. doi:10.2118/98-02-06

Medeiros F et al (2007) Pressure-transient performances of hydraulically fractured horizontal wells in locally and globally naturally fractured formations. Paper presented at the international petroleum technology conference, Dubai, UAE, 4–6 Dec. doi:10.2523/11781-MS

Medeiros F et al (2008) Productivity and drainage area of fracture horizontal wells in tight gas reservoirs. SPE Res Eval Eng 11(5):902–911. doi:10.2118/108110-PA

Mukherjee H et al (1991) Extension of fractured decline curve analysis to fissured formations. Paper presented at the low-permeability reservoirs symposium, Denver, Colorado, 15–17 April 1991. doi:10.2118/21872-MS

Nelson B et al (2014) Predicting long-term production behavior of the Marcellus shale. Paper presented at the SPE western North American and Rocky mountain joint meeting, Denver, Colorado, 17–18 April 2014. doi:10.2118/169489-MS

Nobakht M et al (2012) New type curves for analyzing horizontal well with multiple fractures in shale gas reservoirs. J Nat Gas Sci Eng 10:99–112. doi:10.1016/j.jngse.2012.09.002

Nolte KG (1979) Determination of fracture parameters from fracturing pressure decline. Paper presented at the SPE annual technical conference and exhibition, Las Vegas, Nevada, 23–26 Sept 1979. doi:10.2118/8341-MS

Nolte KG (1986) A general analysis of fracturing pressure decline with application to three models. SPE Form Eval 1(6):571–583. doi:10.2118/12941-PA

Nolte KG (1988) Principles for fracture design based on pressure analysis. SPE Prod Eng 3(1):22–30. doi:10.2118/10911-PA

Nolte KG (1997) Background for after-closure analysis of fracture calibration Tests. Unsolicited companion paper to SPE 38676

Nolte KG et al (1997) After-closure analysis of fracture calibration tests. Paper presented at the SPE annual technical conference and exhibition, San Antonio, Texas, 5–8 Oct 1997

Ozkan E et al (2011) Comparison of fractured-horizontal-well performance in tight sand and shale reservoirs. SPE Res Eval Eng 14(2):248–259. doi:10.2118/121290-PA

Palacio JC, Blasingame TA (1993) Decline-curve analysis using type curves—analysis of gas well production data. Paper presented at the SPE joint Rocky mountain regional and low permeability reservoirs symposium, Denver, Colorado, 26–28 April 1993

Pedrosa OA (1986) Pressure transient response in stress-sensitive formations. Paper presented at the SPE California regional meeting, Oakland, California, 2–4 April 1986. doi:10.2118/15115-MS

Raghavan R, Chin LY (2004) Productivity changes in reservoirs with stress-dependent permeability. SPE Res Eval Eng 7(04):308–315. doi:10.2118/88870-PA

Raghavan RS et al (1997) An analysis of horizontal wells intercepted by multiple fractures. Soc Pet Eng J 2(3):235–245. doi:10.2118/27652-PA

Rana S, Ertekin T (2012) Type curves for pressure transient analysis of composite double-porosity gas reservoirs. Presented at the SPE western regional meeting, Bakersfield, California, 19–23 March 2012. doi:10.2118/153889-MS

Soliman MY et al (2005) After-closure analysis to determine formation permeability, reservoir pressure, and residual fracture properties. Paper presented at the SPE annual technical conference and exhibition, New Orleans, Louisiana, 4–7 Oct 2005. doi:10.2118/124135-MS

Talley GR et al (1999) Field application of after-closure analysis of fracture calibration tests. Paper presented at the SPE mid-continent operations symposium, Oklahoma City, Oklahoma, 28–31 March 1999. doi:10.2118/52220-MS

Tran D et al (2005) An overview of iterative coupling between geomechanical deformation and reservoir flow. Paper presented at the SPE international thermal operations and heavy oil symposium, Calgary, Alberta, 1–3 Nov 2005

Valko PP (2009) Assigning value to stimulation in the Barnett shale: a simultaneous analysis of 7000 plus production histories and well completion records. Paper presented at the SPE hydraulic fracturing technology conference, Woodlands, Texas, 19–21 Jan 2009. doi:10.2118/119369-MS

Valko PP, Lee WJ (2010) A better way to forecast production from unconventional gas wells. Paper presented at the SPE annual technical conference and exhibition, Florence, Italy, 19–22 Sept 2010. doi:10.2118/134231-MS

Wu Y-S et al (2012) Transient pressure analysis of gas wells in unconventional reservoirs. Paper presented at the SPE Saudi Arabia section technical symposium and exhibition, Al-Khobar, Saudi Arabia, 8–11 April 2012. doi:10.2118/160889-MS

Yeager BB, Meyer BR (2010) injection/fall-off testing in the Marcellus shale: using knowledge to improve operational efficiency. Paper presented at the SPE eastern regional meeting, Morgantown, West Virginia, 13–15 Oct 2010. doi:10.2118/139067-MS

Chapter 5
Future Technologies

5.1 Introduction

In the past decade, shale gas resources have received great attention because of their potential to supply the world with an immense amount of energy. However, production of shale gas is small compared with world reserves and it is concentrated in North America. In other to increase the production of shale gas in the entire world, improved technologies are needed. In this chapter, two technologies are introduced. One is CO_2 injection in shale reservoir. Due to rapid decline of gas rate after few years of production, enhanced gas recovery (EGR) technology with CO_2 injection attracts attention. In addition, CO_2 storage in shale reservoir also has received attention because the affinity of CO_2 sorption to the shale reservoir is larger than that of CH_4. The other is advanced well structure. Hydraulic fracturing technique used for shale formations has met with increasing concerns about potential impact on the environment. Fractured wells also show productivity decline due to fracture closure with time and uncertainty of fracture propagation due to the lack of knowledge of formation stresses. Advanced well structures can be a solution for these problems. Advanced well structures are defined as wells having one or more branches tied back to a mother wellbore, which conveys fluids to or from surface. The main advantages of these wells are to increase productivity and reduce development cost.

5.2 CO$_2$ Injection

In past years, the supply of shale resources has increased rapidly throughout the North America and the world. However, in the shale gas well, gas rate decreases rapidly after few years of production. Consequently, interest of enhanced gas

© The Author(s) 2016
K.S. Lee and T.H. Kim, *Integrative Understanding of Shale Gas Reservoirs*,
SpringerBriefs in Applied Sciences and Technology,
DOI 10.1007/978-3-319-29296-0_5

recovery (EGR) for shale gas reservoir is growing recently. In a shale reservoir, methane (CH_4) is adsorbed on surface of the matrix particle or natural fracture face and stored in matrix and fracture pore as free gas (Kang et al. 2011). Several researches showed that the affinity of carbon dioxide (CO_2) sorption to the shale reservoir is larger than that of CH_4 under the subsurface conditions and depending on the thermal maturity of organic materials (Busch et al. 2008; Shi and Durucan 2008). Furthermore, CO_2 injection is important in shale gas reservoir for not only enhanced CH_4 production but also the storage of CO_2. Stronger affinity of CO_2 to the shale reservoir could initiate mechanisms to displace CH_4 existed originally and to adsorb the CO_2 introduced into the shale gas environment. The CO_2 is also could be stored in some portion of the pore volume as non-adsorbed CO_2, especially where hydraulic fracturing has enhanced injectivity.

Although CO_2 injection in shale gas reservoir has not been commercialized yet, several researchers have investigated this subject (Schepers et al. 2009; Godec et al. 2013; Eshkalak et al. 2014; Liu et al. 2013; Fathi and Akkutlu 2013; Jiang et al. 2014; Yu et al. 2014a). Attempts have been made to study the feasibility of CO_2 injection in the Middle and upper Devonian black shale (Schepers et al. 2009). Schepers et al. (2009) described the reservoir modeling and history matching of a Devonian gas shale play in eastern Kentucky and its potential for CO_2 enhanced gas recovery and storage. Well production was history matched by applying an auto-mated process. Finally, several CO_2 injection scenarios with huff-n-puff and con-tinuous injection were reviewed to evaluate the enhanced gas recovery potential and to assess the CO_2 storage capacity of these shale reservoirs. They concluded that the full-field continuous CO_2 injection seems to be of potential success, allowing injection of 300 tons over a period of one and a half month and showing a sig-nificant gain in the recovery. In addition, depending upon the thickness considered, half to the total volume injected is being sequestered. However, the huff-and-puff scenario does not seem to be a good option for that specific reservoir, generating no enhanced gas recovery due to the CO_2 being reproduced very quickly during the puff periods. Even trying longer soaking periods did not seem to improve the recovery.

Liu et al. (2013) focused on CO_2 storage in Devonian and Mississippian New Albany shale gas play in terms of injectivity, storage capacity, sequestration effectiveness, and its impact on CH_4 production. They showed over 95 % of injected CO_2 is effectively sequestered instantaneously with gas adsorption being the dominate storage mechanism. Microscale studies using optical, nuclear, and petrophysical techniques also support the interpretation that gas shales have abundant nano-scale pores in organic matter that allow CO_2 storage through gas adsorption.

Fathi and Akkutlu (2013) presented new mathematical model based on the Maxwell-Stefan formulation for simulations of multi-component transport between CO_2 and CH_4 in shale reservoirs. The approach considered competitive transport and adsorption effect in the organic micropores of the shale during CO$_2$ injection. Focus of their paper was to develop a new triple-porosity single-permeability flow simulation model which is based on a new kinetic approach for the description of gas release from the organic micropores into the inorganic macropores and fractures. It is shown that the surface diffusion of the adsorbed molecules in the micropores is an important mechanism of transport during the CO$_2$ enhanced shale gas recovery since it leads to important counter diffusion and competitive adsorption effects.

Because CO$_2$ injection technique for shale formations is still in its very preliminary stage in spite of these previous researches, study for more accurate simulation model of CO$_2$ injection in shale reservoir is needed. Kim et al. (2015) used gas well data from Barnett Shale in order to model the shale reservoir more accurately. Using this field data, history matching is performed and reservoir and fracture properties of shale gas reservoir for CO$_2$ injection simulation are obtained. Comprehensive reservoir simulation models are presented to investigate effective CO$_2$ injection strategy considering reservoir and fracture properties. Sensitivity analysis for either enhanced CH$_4$ recovery or CO$_2$ storage is conducted to investigate the critical parameters that control CO$_2$-EGR process and CO$_2$ storage, respectively. In following section, this work is introduced in detail, which is important for better understanding of basic mechanisms and proper design of CO$_2$ injection in order to enhance CH$_4$ recovery and CO$_2$ storage.

Although dissolution, residual, and mineral trapping are known as the general trapping mechanisms for immobilization of CO$_2$ in geological media, in shale gas reservoir, adsorption trapping is a dominant mechanism for CO$_2$ storage due to affinity of CO$_2$ to the organic shale. In order to calculate competitive multi-component adsorption/desorption in the model, extended Langmuir isotherm, which has been proven to present a reasonable correlation of the CH$_4$ and CO$_2$ binary gas sorption, is applied given bellow (Arri et al. 1992; Hall et al. 1994),

$$\omega_i = \frac{\omega_{i,\max} B_i y_{ig} p}{1 + p \sum_j B_j y_{jg}}, \tag{5.1}$$

where ω_i is the moles of adsorbed component i per unit mass of rock, $\omega_{i,\max}$ the maximum moles of adsorbed component i per unit mass of rock, B_i the parameter for Langmuir isotherm relation, y_{ig} the molar fraction of adsorbed component i in the gas phase, and p the pressure.

Dissolution trapping is considered by gas solubility represented by Henry's law. Dissolution of the component i in the reservoir fluid is calculated by Henry's law as follows (Li and Nghiem 1986),

$$y_{iw} H_i = f_{iw}, \tag{5.2}$$

where y_{iw} is the mole fraction of component i in the aqueous phase, H_i the Henry's constant of component i, and f_{iw} the fugacity of component i in the aqueous phase.

The aqueous phase and the gaseous phase are assumed in thermodynamic equilibrium so that f_{iw} is equal to the fugacity of component i in the gas phase f_{ig}. f_{ig} is computed from Peng and Robinson (1976) equation of state. The Henry's constants H_i are calculated by below equation (Stumm and Morgan 1996).

$$\ln H_i = \ln H_i^* + \frac{\bar{V}_i(p - p^*)}{RT},$$ (5.3)

where H_i^* is the Henry's constant for component i at reference pressure p^*, $\bar{V}_i$ the partial molar volume of component i, p^* the reference pressure, R the universal gas constant, and T the temperature.

Due to ultra-low permeability of shale matrix, diffusion in the reservoir is significantly important. Especially, when CO_2 is introduced in the reservoir, effect of molecular diffusion between CH_4 and CO_2 should be considered. Sigmund (1976a, b) conducted experiments for various gases to investigate the binary diffusion coefficient. From the results of the experiments, following polynomial was obtained by fitting with the observed values.

$$D_{ij} = \frac{\rho^0 D_{ij}^0}{\rho}\left(0.99589 + 0.096016\rho_r - 0.22035\rho_r^2 + 0.032874\rho_r^3\right),$$ (5.4)

where D_{ij} is the binary diffusion coefficient between component i and j in the mixture, $\rho^0 D_{ij}^0$ the zero pressure limit of the density-diffusivity product, ρ the molar density of the diffusing mixture, and ρ_r the reduced density. From the above equation for binary diffusion coefficient, the diffusion coefficient of component i in the mixture can be computed as follow,

$$D_i = \frac{1 - y_i}{\sum_{j \neq i} y_i D_{ij}^{-1}},$$ (5.5)

where D_i is the diffusion coefficient of component i in the mixture and y_i the mole fraction of component i. With this calculation, competitive diffusion between CH_4 and CO_2 was modeled.

Previous studies showed that linear-elastic model cannot solely describe the geomechanical effects of shale gas reservoirs (Li and Ghassemi 2012; Hosseini 2013). Meanwhile, in order to consider the change of reservoir conductivity, pressure-dependent permeability was presented in several researches (Pedrosa 1986; Raghavan and Chin 2004; Cho et al. 2013). Therefore, the deformation of shale reservoir is modeled by stress-dependent correlations coupled with linear-elastic model. Exponential correlation (Eqs. 2.36 and 2.37) is used to calculate these stress-dependent porosity and permeability. Experimental coefficients are obtained from Cho et al. (2013).

In order to analyze the realistic effects of CO_2 injection in shale gas reservoir, numerical model of shale gas reservoir was generated based on properties from history matching. Field data of Barnett Shale reproduced from Anderson

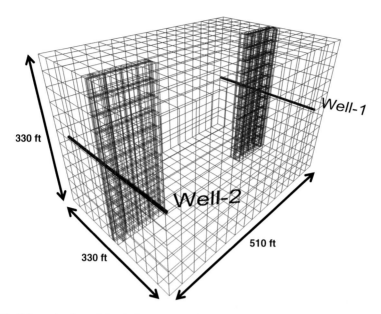

Fig. 5.1 Schematic view of the shale gas reservoir model for the CO_2 injection

et al. (2010) were used to perform history matching. Using the results of history matching, the segment of reservoir, which is simplified for computational efficiency, was generated with two hydraulically fractured horizontal well (Fig. 5.1). The size of the segment is $330 \times 510 \times 330$ ft^3. In this model, dual porosity dual permeability model was considered to characterize the matrix and natural fracture system in shale gas reservoir. The horizontal wells are located at the center of the reservoir and hydraulic fractures are located at the center of the each well. Local grid refinement (LGR) technique is used to model the thin block assigned with the properties of hydraulic fracture. Height of the hydraulic fractures is same as net pay which shows fully penetrated reservoir. It is assumed that properties of hydraulic fracture are constant along the fracture and have finite conductivity. To analyze the effects of the geomechanical model, exponential correlation is considered coupled with linear-elastic model. In the simulation scenario, at first, two horizontal wells are produced for five years. Then, in the well 2, CO_2 is injected while well 1 continues to produce. After five years, CO_2 injector well 2 is shut-in and well 1 is produced for 40 years.

In order to investigate the effects of CO_2 injection for EGR, gas recovery from the model with and without CO_2 injection is presented in Fig. 5.2. This graph shows that recovery of each model with and without CO_2 injection is 51.1 and 38.7 % so that increase of recovery caused by CO_2 flooding is 12.4 % at the end of the production. Figures 5.3 and 5.4 show cumulative gas moles and gas mole rate of CH_4 and CO_2 with and without CO_2 injection observed in the production well 1. Figure 5.3 indicates that about 98 % of the produced gas in the well 1 is CH_4 and

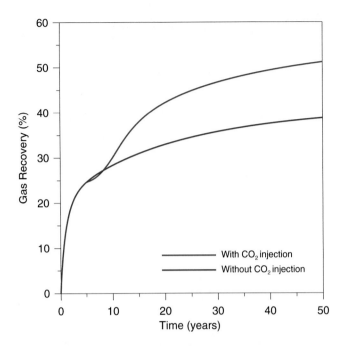

Fig. 5.2 Gas recovery for the model with and without CO_2 injection

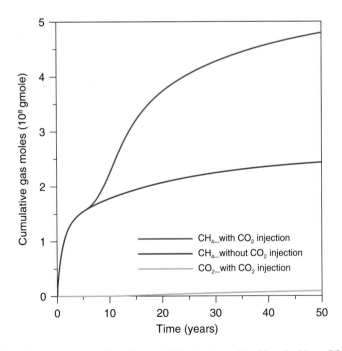

Fig. 5.3 Cumulative gas moles of the CH_4 and CO_2 for the model with and without CO_2 injection

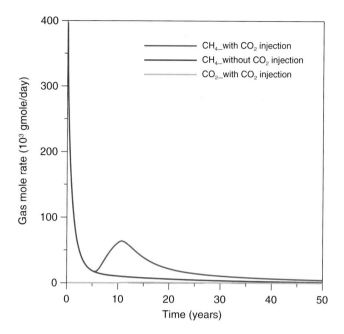

Fig. 5.4 Gas mole rate of the CH$_4$ and CO$_2$ for the model with and without CO$_2$ injection

only about 2 % of the produced gas is CO$_2$. Increasing CH$_4$ production caused by CO$_2$ injection is observed approximately 1 year after injection started. CO$_2$ breakthrough is observed 3 years after injection started in the well 1 but production rate remains extremely low. In Fig. 5.4, the peak of CH$_4$ is shown approximately 1 year after the shut-in and it is 5 times the CH$_4$ rate of model without CO$_2$ injection.

Figure 5.5 shows moles of injected, stored, and produced CO$_2$ in the shale reservoir. In this figure, about 96 % of total injected CO$_2$ is stored in the reservoir and only about 4 % is produced from the production well at the end of the production. Figure 5.6 provides the classification of injected CO$_2$ in the shale reservoir based on the states of CO$_2$ such as super-critical, adsorbed, dissolved, and produced CO$_2$. The CO$_2$ stored in the reservoir exists as a super-critical phase, adsorption trapping, and dissolution trapping of 45.8, 46.5, and 3.6 %, respectively at the end of the production. Among these states of CO$_2$, super-critical phase is mobile but adsorption and dissolution states are immobile because they are trapped in the surface of matrix and connate water respectively. Directly after the end of CO$_2$ injection, the amount of super-critical CO$_2$ is higher than that of adsorbed CO$_2$. However, as time goes on, super-critical CO$_2$ spreads to the reservoir and amount of super-critical CO$_2$ decreases due to increase of adsorption and dissolution trapping.

Figure 5.7 presents adsorbed moles of the CH$_4$ and the CO$_2$ with and without considering the multi-component adsorption mechanism. Solid and dashed lines indicate the amount of CH$_4$ and CO$_2$ adsorption, respectively. Red and blue lines

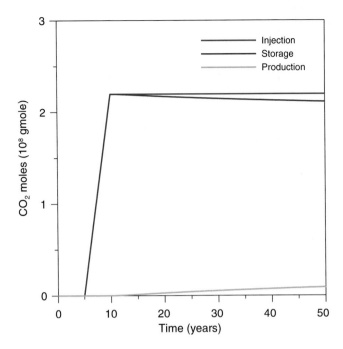

Fig. 5.5 Moles of injected, stored, and produced CO_2

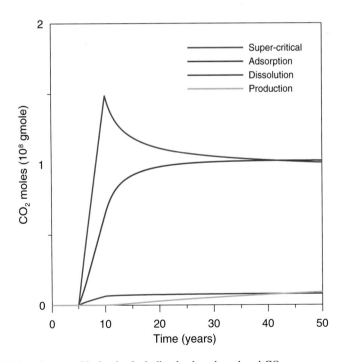

Fig. 5.6 Moles of super-critical, adsorbed, dissolved, and produced CO_2

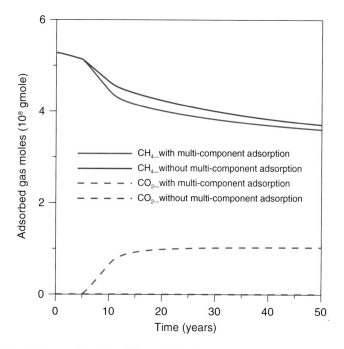

Fig. 5.7 Adsorbed gas moles of the CH$_4$ and CO$_2$ for the model with and without considering multi-component adsorption

indicate whether the multi-component adsorption is considered or not. In the model without the multi-component adsorption, the CO$_2$ injection only causes the effect of pressurizing the reservoir and the CO$_2$ is not adsorbed in the shale reservoir. On the other hand, in the model with the multi-component adsorption, desorption of the CH$_4$ is activated by competitive sorption with the CO$_2$ which is preferentially adsorbed over CH$_4$ with a ratio up to 5:1 based on laboratory and theoretical calculations (Nuttall 2010). In Fig. 5.7, desorption of CH$_4$ is higher in the model with the multi-component adsorption. Figure 5.8 shows that schematic view of the CO$_2$ adsorption in the reservoir. Figure 5.8a through c indicate gmole per cubic feet of CO$_2$ adsorption after 10, 30, 50 years, respectively. After injection stops, while production continues for 40 years, the CO$_2$ migrates to the production well so that the adsorption of CO$_2$ spreads to the reservoir.

Mole fraction of CH$_4$ and CO$_2$ is compared in the model with and without mechanism of the molecular diffusion (Fig. 5.9). These values are measured from the point A of the reservoir model shown in Fig. 5.10. In Fig. 5.9, solid and dashed lines indicate the mole fraction of CH$_4$ and CO$_2$ and red and blue lines indicate whether the molecular diffusion is considered or not. Figure 5.9 shows that the mole fraction of CH$_4$ decreases about 10 % and that of CO$_2$ increases about 10 % due to the effect of molecular diffusion computed by Sigmund correlation (1976a, b). Figure 5.10 also presents schematic view of the CH$_4$ mol fraction in the reservoir at

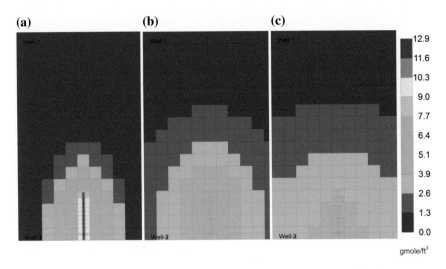

Fig. 5.8 Schematic views of the adsorbed CO_2 distribution at **a** 10, **b** 30, and **c** 50 years

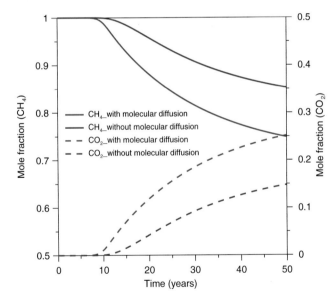

Fig. 5.9 Mole fraction of the CH_4 and CO_2 for the model with and without considering molecular diffusion

the end of the production. Figure 5.10a which shows the model considering molecular diffusion presents more widely-spread CO_2 in the reservoir compared with Fig. 5.10b. Because of ultra-low permeability of shale reservoir, effect of diffusion is higher than conventional reservoirs so that it should be considered in the CO_2 injection model of shale reservoir.

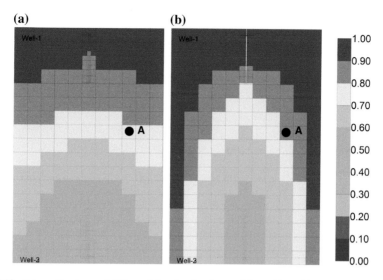

Fig. 5.10 Schematic views of CH_4 mol fraction for the model **a** with and **b** without considering molecular diffusion at the end of the production

Geomechanical effects are also important in the shale reservoir (Cho et al. 2013). In order to calculate the geomechanical effects, stress-dependent porosity and permeability coupled with linear-elastic model is applied in this model. Exponential correlations are used to compute the stress-dependent properties. Figure 5.11 shows the change of natural fracture permeability in the model with and without CO_2 injection. In first five years, permeability decreases rapidly due to the decrease of pressure during production. After injection begins, natural fracture permeability increases until the shut-in of injection well and decreases again. Due to the pressurizing effect caused by injection of CO_2, average pressure increases shown as Fig. 5.12 so that permeability increases depending on the stress-dependent correlation. Increment of porosity and permeability caused by geomechanical model has a positive effect on the CO_2 injection in the shale reservoir.

Shale gas reservoirs show high uncertainty because of inestimable reservoir and fracture properties. Although several sensitivity studies for CO_2-EGR were performed (Kalantari-Dahaghi 2010; Jiang et al. 2014; Yu et al. 2014a), there are no results for the CO_2 storage in shale formations. In this study, sensitivity analysis was performed for the both CH_4 recovery and the CO_2 storage. Especially, in stored CO_2, adsorption and dissolution states are immobile because they are trapped in the surface of matrix and connate water and they are important for the CO_2 storage due to stability. Therefore, sensitivity analysis for the trapped CO_2 is also conducted. Figures 5.13, 5.14 and 5.15 provide results of sensitivity analysis for the objective functions of cumulative produced moles of CH_4, stored moles of CO_2 in reservoir, and trapped moles of CO_2. In this study, uncertain parameters considered for sensitivity analysis are porosity of matrix and natural fractures, permeability of

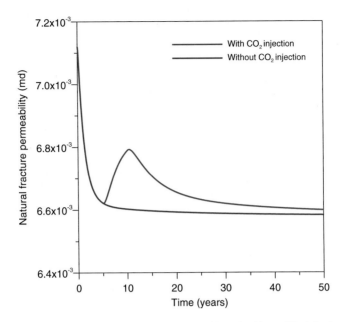

Fig. 5.11 Natural fracture permeability for the model with and without CO_2 injection

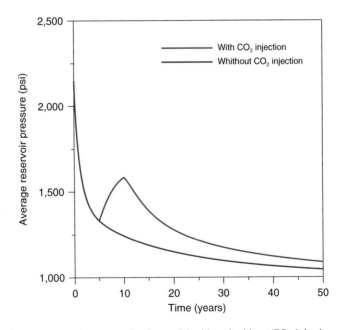

Fig. 5.12 Average reservoir pressure for the model with and without CO_2 injection

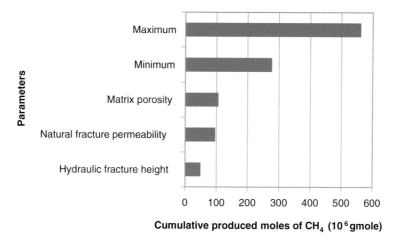

Fig. 5.13 Result of sensitivity analysis for enhanced CH₄ recovery

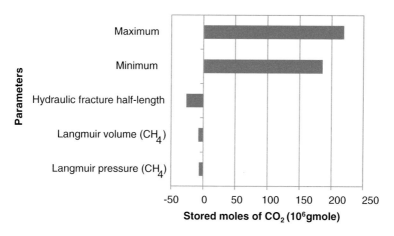

Fig. 5.14 Result of sensitivity analysis for CO₂ storage

matrix, natural fractures, and hydraulic fractures, hydraulic fracture height, hydraulic fracture half-length, Langmuir volume, and Langmuir pressure. Effects of parameters which show high influence to the each sensitivity analysis were presented in Figs. 5.13, 5.14 and 5.15. For the CO₂-EGR, matrix porosity, natural fracture permeability and height are significantly of importance. It show that effects of EGR increase with high fracture conductivity. On the other hand, for the CO₂ storage, influence of uncertain parameters are small compared with the result of EGR. In this case, hydraulic fracture half-length is most important and dominant parameter. For the CO₂ trapping, Langmuir constants are main parameters.

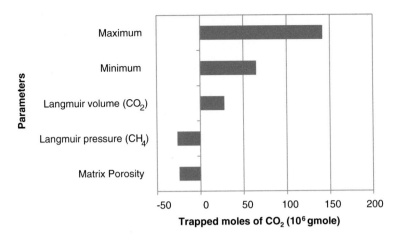

Fig. 5.15 Result of sensitivity analysis for trapped CO_2

Therefore, these parameters should be mainly considered to inject CO_2 in the shale gas reservoir for each objective.

Recently, enhanced oil recovery (EOR) in shale oil reservoir is also concentrated as well as EGR in shale gas reservoir. Tovar et al. (2014) presented experimental results on the use of CO_2 as an EOR agent in preserved, rotary sidewall reservoir core samples with negligible permeability. The results of this investigation support CO_2 as a promising EOR agent for shale oil reservoir. Oil recovery was estimated to be between 18 and 55 % of OOIP. They provided a detailed description of the experimental set up and procedures. The analysis of the x-ray computed tomography images revealed saturation changes within the shale core as a result of CO2 injection. Chen et al. (2014), Yu et al. (2014b), Wan and Sheng (2015) presented simulation study for EOR in shale oil reservoir. In spite of these previous studies, research of EGR and EOR in shale reservoir is still lacking. In the future, for stable production in shale reservoirs, more research of this area is needed.

5.3 Advanced Well Structure

Even though it is generally agreed that a vast amount of resources are locked within these unconventional systems, the challenges for effective and economical production still seem daunting for the current state of technology. This is tied to the dearth of in-depth knowledge about the complexities of these systems and the lack of mathematical and analytical techniques that adequately capture them. Existing production techniques such as hydraulic fracturing, which is heavily used for shale and tight formations, have met with increasing concerns about their potential impact on the environment (Enyioha and Ertekin 2014). In addition, this type of wells show productivity decline due to fracture closure with time and uncertainty of fracture propagation due to the lack of knowledge of formation stresses. Therefore, extensive effort has been directed towards developing effective alternatives for field development tools.

Advanced well structures (or multi-lateral wells) are defined as wells having one or more branches (laterals) tied back to a mother wellbore, which conveys fluids to or from surface. Technology of horizontal and multilateral fishbone wells provides significant leverage where conventional vertical wells cannot efficiently maintain a profitable development (Bukhamseen 2014). The main advantages of these well configurations are to increase well productivity and reduce development cost per field. Multi-lateral wells can produce higher rates of oil and gas because they have a larger reservoir contact area compared to vertical wells (Charlez and Breant 1999). Not only these wells produce more, but they also provide better sweep efficiency by mitigating or preventing gas and water conings as the position of the laterals within the producing layers provides enough distance to water and gas bearing zones. In addition, more reserves are realized due to the extended reach of multilateral wells which creates a larger drainage area. Consequently, a large field can be developed with less number of wells and therefore, multilateral wells can reduce time and costs of drilling and surface facilities construction operations (Ismail and El-Khatib 1996). These well structures also reduce surface footprints by eliminating the need for multiple drilling pads whilst still effecting adequate surface area contact with the reservoir.

Advanced well structures have seen increased field application over the years. Joshi (2000) reported that over 700 multi-lateral wells have been drilled in Saskatchewan, Canada. Making a comparison with single lateral wells, Stalder et al. (2001) report that advanced well structures yield higher recovery factors because of the extra leverage gained from the ability to produce from multiple targets, thus sustaining the declining rates and keeping the well operationally feasible for longer periods. The various forms of multilateral wells such as stacked dual lateral, gullwing multi-lateral, crow's foot triple lateral, pitchfork dual lateral, and fishbone wells are drilled. Figure 5.16 shows two adjacent pad developed with a combination of stacked dual lateral, gullwing, crow's foot, and fishbone multilateral wells.

Advanced well structures have also been deployed in the Shaybah field of Saudi Arabia (Saleri et al. 2004). Fish bone well (SHYB-220) was drilled with a total of eight laterals comprising an aggregate reservoir contact of 12.3 km as part of a pilot program (Fig. 5.17). A production test on SHYB-220 indicated a PI of 126 STB/D/psi, which represents a six-fold increase compared to 1-km horizontal completions in similar facies (10 md). Furthermore, a four-fold reduction in unit-development costs was achieved. These well systems are particularly useful in producing from thin formations, formations with isolated pockets of producing zones, and unconventional reservoirs. Advanced well structures compare favorably with other well design options.

Yu et al. (2009) showed close comparison in net present value from development plans using fish bone well and hydraulic fractures. They founded that these two types of wells can generate comparable NPV values. They concluded that as the drilling technology develops, use of fishbone wells with increased number of rib

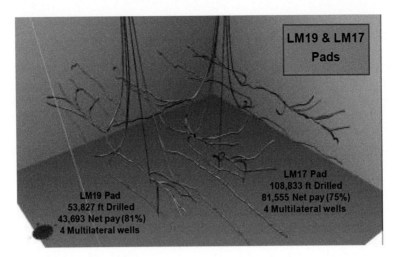

Fig. 5.16 Two adjacent pad developed with a combination of stacked dual lateral, gullwing, crow's foot, and fishbone multilateral wells (Stalder et al. 2001)

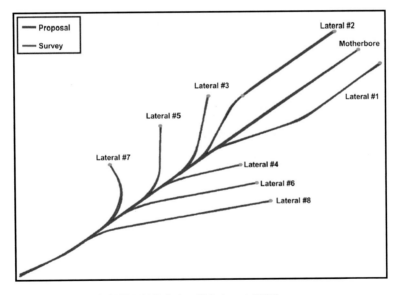

Fig. 5.17 Fish bone well (SHYB-220) design (Saleri et al. 2004)

holes will be more beneficial than multi-fractured wells in developing tight gas fields.

Enyioha and Ertekin (2014) performed simulation study for advanced well structure model to forecast production performance of unconventional reservoirs. Enyioha and Ertekin (2014) proposed a set of forward-acting and inverse-acting

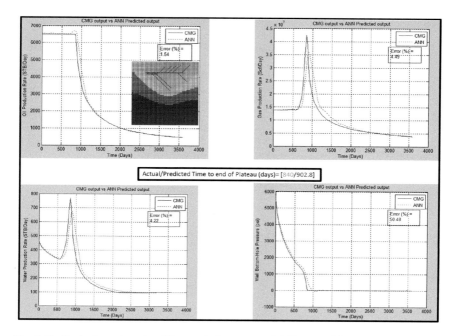

Fig. 5.18 Production rate and well bottom hole pressure of fishbone well in numerical model and artificial neural network model (Enyioha and Ertekin 2014)

predictor models based on an artificial neural network framework and applied to advanced well structures in tight multi-phase systems. The forward-acting models forecast production rates, while the inverse-acting models generate well designs that can meet desired cumulative production profiles. Figure 5.18 shows plots of oil rate, water rate, gas rate, and wellbore flowing pressure of fishbone well in numerical model and artificial neural network model.

References

Anderson D et al (2010) Analysis of production data from fractured shale gas wells. Soc Pet Eng J 15(01):64–75. doi:10.2118/115514-PA

Arri LE et al (1992) Modeling coalbed methane production with binary gas sorption. Paper presented SPE rocky regional meeting. Casper, Wyoming, 18–21 May 1992. doi:10.2118/24363-MS

Bukhamseen I (2014) Artificial expert systems for rate transient analysis of fishbone wells completed in shale gas reservoirs. Dissertation, The Pennsylvania State University, USA

Busch A et al (2008) Carbon dioxide storage potential of shales. Int J Greenh Gas Con 2(3):297–308. doi:10.1016/j.ijggc.2008.03.003

Charlez PA, Breant P (1999) The multiple role of unconventional drilling technologies. Paper presented at the SPE European formation damage conference June. The Hague, Netherlands, 31 May–1 1999. doi:10.2118/56405-MS

Chen C et al (2014) Effect of reservoir heterogeneity on primary recovery and CO_2 huff 'n' puff recovery in shale-oil reservoirs. SPE Res Eval Eng 17(3):404–413. doi:10.2118/164553-PA

Cho Y et al (2013) Pressure-dependent natural-fracture permeability in shale and its effect on shale-gas well production. SPE Res Eval Eng 16(2):216–228. doi:10.2118/159801-PA

Enyioha C, Ertekin T (2014) Advanced well structures: an artificial intelligence approach to field deployment and performance prediction. Paper presented at the SPE intelligent energy conference & exhibition. Utrecht, The Netherlands, 1–3 April. doi:10.2118/167870-MS

Eshkalak MO et al (2014) Enhanced gas recovery by CO_2 sequestration versus re-fracturing treatment in unconventional shale gas reservoirs. Paper presented at the Abu Dhabi international petroleum exhibition and conference. Abu Dhabi, UAE, 10–13 Nov 2014. doi:10.2118/172083-MS

Fathi E, Akkutlu IY (2013) Multi-component gas transport and adsorption effects during CO_2 injection and enhanced shale gas recovery. Int J Coal Geol 123:52–61. doi:10.1016/j.coal. 2013.07.021

Godec M et al (2013) Potential for enhanced gas recovery and CO_2 storage in the Marcellus shale in the Eastern United States. Int J Coal Geol 118:95–104. doi:10.1016/j.coal.2013.05.007

Hall FE et al (1994) Adsorption of pure methane, nitrogen, and carbon dioxide and their binary mixtures on wet Fruitland coal. Paper presented at the 1994 eastern regional conference & exhibition. Charleston, West Virginia, 8–10 Nov. doi:10.2118/29194-MS

Hosseini SM (2013) On the linear elastic fracture mechanics application in Barnett shale hydraulic fracturing. Presented at the 47th U.S. rock mechanics/geomechanics symposium. San Francisco, California, 23–26 June 2013

Ismail G, El-Khatib H (1996) Multilateral horizontal drilling problems & solutions experiences offshore. Paper presented at the Abu Dhabi international petroleum exhibition and conference. Abu Dhabi, UAE, 13–16 Oct 1996. doi:10.2118/36252-MS

Jiang J et al (2014) Development of a multi-continuum multi-component model for enhanced gas recovery and CO_2 storage in fractured shale gas reservoirs. Paper presented at the SPE improved oil recovery symposium, 12–16 April. Tulsa, Oklahoma. doi:10.2118/169114-MS

Joshi S (2000) Horizontal and multi-lateral wells: performance analysis—an art or a science? J Can Pet Tech 39(10):19–23

Kalantari-Dahaghi A (2010) Numerical simulation and modeling of enhanced gas recovery and CO_2 sequestration in shale gas reservoirs: a Feasibility Study. Paper presented at the SPE international conference on CO_2 capture, storage, and utilization. New Orleans, Louisiana, 10–12 Nov 2010. doi:10.2118/139701-MS

Kang SM et al (2011) Carbon dioxide storage capacity of organic-rich shales. Soc Pet Eng J 16 (04):842–855. doi:10.2118/134583-PA

Kim TH et al (2015) Modeling of CO_2 injection considering multi-component transport and geomechanical effect in shale gas reservoirs. Paper presented at the SPE/IATMI Asia Pacific oil & gas conference and exhibition, Bali, 20–22 Oct 2015. http://dx.doi.org/10.2118/176174-MS

Li Y, Ghassemi A (2012) Creep behavior of Barnett, Haynesville, and Marcellus shale. Paper presented at the 46th U.S. rock mechanics/geomechanics symposium. Chicago, Ilinois. 24–27 June 2012

Li YK, Nghiem LX (1986) Phase equilibria of oil, gas and water/brine mixtures from a cubic equation of state and Henry's law. Can J Chem Eng 64(3):486–496. doi:10.1002/cjce. 5450640319

Liu F et al (2013) Assessing the feasibility of CO_2 storage in the new Albany Shale (Devonian–Mississippian) with potential enhanced gas recovery using reservoir simulation. Int J Greenh Gas Cont 17:111–126. doi:10.1016/j.ijggc.2013.04.018

Nuttall BC (2010) Reassessment of CO_2 sequestration capacity and enhanced gas recovery potential of middle and upper Devonian black shales in the Appalachian basin. In: MRCSP Phase II topical report, Kentucky geological survey. Lexington, Kentucky, 2005 October–2010 October

Pedrosa OA (1986) Pressure transient response in stress-sensitive formations. Paper presented at the SPE California regional meeting. Oakland, California, 2–4 April 1986. doi:10.2118/15115-MS

Peng DY, Robinson DB (1976) A new two-constant equation of state. Ind Eng Chem Fund 15(1):59–64. doi:10.1021/i160057a011

Raghavan R, Chin LY (2004) Productivity changes in reservoirs with stress-dependent permeability. SPE Res Eval Eng 7(04):308–315. doi:10.2118/88870-PA

Saleri NG et al (2004) Shaybah-220: a maximum-reservoir-contact (MRC) well and its implications for developing tight-facies reservoirs. SPE Res Eval Eng 7(4):316–321. doi:10.2118/81487-MS

Schepers KC et al (2009) Reservoir modeling and simulation of the Devonian gas shale of Eastern Kentucky for enhanced gas recovery and CO_2 storage. Paper presented at the SPE international conference on CO_2 capture, storage, and utilization. San Diego, California, 2–4 Nov. doi:10.2118/126620-MS

Shi JQ, Durucan S (2008) Modeling of mixed-gas adsorption and diffusion in coalbed reservoirs. Paper presented at the SPE unconventional reservoirs conference. Dallas, Texas, 10–12 Feb 2008. doi:10.2118/114197-MS

Sigmund PM (1976a) Prediction of molecular diffusion at the reservoir conditions, part I-measurement and prediction of binary dense gas diffusion coefficients. J Can Pet Tech 15(2):48–57. doi:10.2118/76-02-05

Sigmund PM (1976b) Prediction of molecular diffusion at the reservoir conditions, part II–estimating the effects of molecular diffusion and convective mixing in multicomponent systems. J Can Pet Tech 15(3):53–62. doi:10.2118/76-03-07

Stalder JL et al (2001) Multilateral-horizontal wells increase rate and lower cost per barrel in the Zuata field, Faja, Venezuela. Paper presented at the SPE international thermal operations and heavy oil symposium. Porlamar, Margarita Island, Venezuela, 12–14 Mar 2001. doi:10.2118/69700-MS

Stumm W, Morgan JJ (1996) Aquatic chemistry: chemical equilibria and rates in natural waters, 3rd edn. Wiley, New York

Tovar FD et al (2014) Experimental investigation of enhanced recovery in unconventional liquid reservoirs using CO_2: a look ahead to the future of unconventional EOR. Paper presented at the SPE international conference on CO_2 capture, storage, and utilization, 10–12 Nov. New Orleans, Louisiana. doi:10.2118/139701-MS

Wan T, Sheng J (2015) Compositional modelling of the diffusion effect on EOR process in fractured shale-oil reservoirs by gasflooding. J Can Pet Tech 54(2):107–115. doi:10.2118/2014-1891403-PA

Yu W et al (2014a) A sensitivity study of potential CO_2 injection for enhanced gas recovery in Barnett shale reservoirs. Paper presented at the SPE unconventional resources conference, Woodlands, Texas, 1–3 Apr 2014. doi:10.2118/169012-MS

Yu W et al (2014b) Simulation study of CO_2 huff-n-puff process in Bakken tight oil reservoirs. Paper presented at the SPE western north American and Rocky mountain joint meeting. Denver, Colorado, 17–18 Apr 2014. doi:10.2118/169575-MS

Yu X et al (2009) A comparison between multi-fractured horizontal and fishbone wells for development of low-permeability fields. Paper presented at the Asia Pacific oil and gas conference & exhibition. Jakarta, Indonesia, 4–6 Aug 2009. doi:10.2118/120579-MS